# DEVELOPMENTS IN PLASTICS TECHNOLOGY—4

# CONTENTS OF VOLUMES 2 AND 3

## Volume 2

## Volume 3

# DEVELOPMENTS IN PLASTICS TECHNOLOGY—4

*Edited by*

## A. WHELAN and J. P. GOFF

*London School of Polymer Technology,*
*The Polytechnic of North London, Holloway, London, UK*

## ELSEVIER APPLIED SCIENCE
LONDON and NEW YORK

ELSEVIER SCIENCE PUBLISHERS LTD
Crown House, Linton Road, Barking, Essex IG11 8JU, England

*Sole Distributor in the USA and Canada*
ELSEVIER SCIENCE PUBLISHING CO., INC.
655 Avenue of the Americas, New York, NY 10010, USA

WITH 40 TABLES AND 172 ILLUSTRATIONS

© 1989 ELSEVIER SCIENCE PUBLISHERS LTD
© 1989 VICKERS SYSTEMS LTD—Chapter 7

**British Library Cataloguing in Publication Data**

Developments in plastics technology.
4
1. Plastics. processing
I. Whelan, A. (Anthony), *1937–*
II. Goff, J. P.
668.4

**Library of Congress Cataloguing in Publication Data**

**Developments in plastics technology.**—1——London; New York
: Applied Science Publishers, c1982–
v.:ill.; 23 cm.—(Developments series)

1. Plastics—Collected works.   I. Series.
TP1101.D48          668.4′05—dc19          84-644560

ISBN 1-85166-268-5

Printed in Great Britain by Galliard (Printers) Ltd, Great Yarmouth

# PREFACE

Because of the sheer size and scope of the plastics industry, the title *Developments in Plastics Technology* now covers an incredibly wide range of subjects or topics. No single volume can survey the whole field in any depth and what follows is, therefore, a series of chapters on selected topics. The topics were selected by us, the editors, because of their immediate relevance to the plastics industry.

When one considers the advancements of the plastics processing machinery (in terms of its speed of operation and conciseness of control), it was felt that several chapters should be included which related to the types of control systems used and the correct usage of hydraulics.

The importance of using cellular, rubber-modified and engineering-type plastics has had a major impact on the plastics industry and therefore a chapter on each of these subjects has been included.

The two remaining chapters are on the characterisation and behaviour of polymer structures, both subjects again being of current academic or industrial interest. Each of the contributions was written by a specialist in that field and to them all, we, the editors, extend our heartfelt thanks, as writing a contribution for a book such as this, while doing a full-time job, is no easy task.

A. WHELAN
J. P. GOFF

# CONTENTS

# LIST OF CONTRIBUTORS

H. BRIEM

*Vickers Systems GmbH, Bad Homburg, West Germany*

J. A. BRYDSON

*Brett Lodge, Brent Eleigh, Sudbury, Suffolk CO10 9NR, UK*

K. T. COLLINGTON

*Schering Industrial Chemicals, Mount Pleasant House, Huntingdon Road, Cambridge CB3 0DA, UK*

R. DIAZ CALLEJA

*Universidad Politecnica, Escuela Tecnica Superior de Ingenieros Industriales, Valencia, Spain*

J. L. GOMEZ RIBELLES

*Universidad Politecnica, Escuela Tecnica Superior de Ingenieros Industriales, Valencia, Spain*

ix

P. W. MILNER

*Goodyear Chémicals Europe, European Chemical Division, Avenue des Tropiques, Z. A. Courtaboeuf, BP 31, 91941 Les Ulis Cedex, France*

KEITH T. O'BRIEN

*Celanese Engineering Resins, 26 Main Street, Chatham, NJ 07928, USA.* Present address: *New Jersey Institute of Technology, University Heights, Newark, New Jersey 07102, USA*

S. SKINNER

*Vickers Systems Ltd, PO Box 4, New Lane, Havant, Hampshire PO9 2NB, UK*

MASAO TAKAHASHI

*Toray Techno Co. Ltd, 1-1 Sonoyama 1-Chome, Otsu 520, Japan*

*Chapter 1*

# PROCESSING OF CELLULAR THERMOPLASTICS

K. T. COLLINGTON

*Schering Industrial Chemicals, Cambridge, UK*

## 1 INTRODUCTION

The processing of cellular thermoplastics now constitutes an established, definable sector within the conventional thermoplastics processing industry. Cellular thermoplastics can be produced by use of the most established thermoplastics processing techniques. The injection moulding and extrusion of cellular thermoplastics have been the subject of an ongoing development intended both to reduce density and (probably of greater importance) to improve surface quality, eliminating or minimising surface finishing operations. Other areas where development in processing is apparent include rotational moulding and press moulding techniques.

Improvements in process development, together with relevant advances in expansion systems and polymer systems used, are outlined in this chapter together with possible areas of future development.

Developments in expandable PVC plastisols are not discussed. Whilst it is realised that the area of coated fabrics and other substrates used in the production of cellular floorcovering and wallcovering, etc., constitutes a major area of end application for mainly chemical expansion systems, it is felt that the technology differs considerably from that discussed in the chapter and it has therefore been excluded.

## 2 INJECTION MOULDING

The processing techniques used in the production of cellular thermoplastic mouldings have been the subject of an ongoing development programme since the original concept in 1962.[1]

The original concept of foam moulding was the development of a low-cost moulding technique for the production of very large mouldings with a high projected surface area. Advantages of the process include low mould clamping forces, with a high degree of design flexibility in the mouldings, giving the mouldings distinct cost advantages over homogeneous mouldings of similar weight and configuration.

Initial applications for such cellular mouldings were in the wood replacement areas, i.e. furniture parts and audio equipment, where full advantage of the low-cost processing technique could be utilised. However, changes in the performance requirements of such mouldings, if they were to compete with homogeneous mouldings in semi-metal replacement areas and at a later stage in metal replacements, quickly dictated changes in both processing techniques and polymer systems.

The transition from wood to metal replacement applications posed immediate problems such as

(1)  the need to improve the surface quality of the mouldings by improving moulding techniques to minimise 'out-of-mould' finishing, or to improve surface finishing techniques;

(2)  the need to improve and maintain the physical properties of consecutive mouldings (particularly impact and deflection loading characteristics).

Details of the sequence of developments in cellular moulding techniques are shown in Table 1. Details of the development of the individual moulding techniques, the polymer systems utilised, and the types of expansion systems, will be discussed individually. Where applicable the process techniques are compared to enable assessments of possible future developments, in the injection moulding of cellular thermoplastics, to be made.

## 2.1 Process Development

Table 1 lists the theoretical stages in the development of areas of application for cellular mouldings and highlights the increasing complexity of the foam moulding process. The pattern of development outlined is *theoretical*. Whilst progression from wood replacement to semi-metal application areas has occurred, progress on the use of cellular mouldings in true metal replacement applications remains limited; this is probably attributable to the lack of adequate engineering design data and doubts on the reproducibility of physical properties.

TABLE 1
SEQUENCE OF DEVELOPMENTS IN THE MOULDING OF CELLULAR THERMOPLASTICS

| Foam classification | Performance requirements of the moulding | Polymers currently utilised | Additives present in the polymer | Expansion system | Moulding system |
|---|---|---|---|---|---|
| (1) Wood replacement | Smooth surface Stain-free surface Uniform fine cell structure | Impact polystyrene Polyphenylene oxide/styrene alloys | Flame retardants Impact modifiers Pigments | Physical Chemical | Low-pressure |
| (2) Semi-metal replacement | As above, but improved surface quality | Polypropylene ABS Polyphenylene oxide/styrene alloys HIPS | Flame retardants Reinforcing fillers Reinforcing fibres UV stabilisers Pigments | Chemical Physical | Low-pressure High-pressure Counter-pressure Integral skin |
| (3) Metal replacement | As (2) above | Polycarbonate Thermoplastic polyesters Polyamides ABS Polyether sulphone Polypropylene | As (2) above | Chemical | As (2) above |
| (4) Speciality foam systems | Dependent on end application, i.e. resilience, density levels, etc. | Composite polymer structures based on polyolefins, or styrenics | As (2) above + crosslinking systems, bonding additives | Chemical | As (2) above + specialist moulding systems |

The reasons for the changes in application areas are essentially economic, resulting from:

(1) the virtual collapse of the audio/TV housings market during the 1976–1980 recession—essentially a wood replacement market;
(2) the emergence of a market for cellular mouldings in housings for computers, business machines, etc.—essentially semi-metal replacements, where dimensional stability, high surface quality, non-flammability and electrical insulation characteristics are essential;
(3) the improved economics of processing polymeric materials by comparison with metal stamping or casting techniques and associated metal finishing costs.[2]

Whilst the pattern of application is similar in most geographic areas, a considerable market for wood replacement applications still exists in North America in the area of reproduction antique furniture.

Individual cellular moulding techniques outlined in Table 2 are, for convenience, divided into two groups—the low-pressure and high-pressure moulding techniques, (independently of the polymer and expansion technique used).

## 2.2 Low-Pressure Moulding Systems

The term 'low-pressure moulding' describes a moulding technique in which a highly vented mould cavity is partially filled by the high-speed injection of a polymer melt in which a dissolved gas is held in solution. Immediately after injection and the subsequent decay of melt pressure to a level below the solution pressure of the dissolved gas, the polymer begins to foam, expanding the polymer melt to fill the mould cavity completely.

This process with slight modifications still remains the most widely used technique for the production of cellular thermoplastic mouldings today. Whilst the mechanics of the process have been adequately documented,[3,4] the two major moulding techniques are outlined below.

### 2.2.1 Conventional Low-Pressure Moulding

This process, based on the above principles, is the most widely established system currently used and will probably remain so in the future. Whilst the mouldings produced are suitable for use in a wide range of end applications, the increasing demands for improved surface finishes and more uniform physical properties will result in little further mechanical development in the process. Additional development is apparent, however, in the area of physical expansion systems where the use of both fluorocarbons and hydrogen[5,6] are reported to give improved cycle times and smooth surfaces. Improvements in surface quality by the thermal cycling of the mould have been claimed,[7] but long cycle times and mould design problems appear to make this approach uneconomic and no commercial operations are thought to exist.

*2.2.1.1 Advantages of conventional low-pressure moulding.* The obvious advantages of the process include the following.

(1) Minimal modifications to the standard 'in-line' moulding equipment, e.g. boosted injection speeds and positive displacement shut-off units (nozzles), permit its use for the production of cellular mouldings. However, the use of a screw pre-plasticising ram injection machine is essential for large mouldings and the economic utilisation of clamp capacities.

(2) The low clamping forces involved still permit the production of mouldings with a large shot weight and projected surface area.

(3) Low internal mould pressures (25–30 psi; $0.2 \, \text{MN m}^{-2}$) permit the use of low-cost steel tooling with, if necessary, aluminium inserts to improve heat extraction in an inherently long cooling cycle.[8]

*2.2.1.2 Disadvantages of conventional low-pressure moulding.* These advantages are offset to some degree by the following aspects.

(1) The reduced clamping capacity precludes the use of the equipment for the production of large homogeneous mouldings, although recently designed equipment has increased clamping capacity to permit dual usage.

(2) Within the limitations of density reduction, moulded parts will require 'out-of-mould' secondary finishing operations, particularly in semi-metal replacement applications.

(3) Increased cycle times attributable to the long cooling cycle require careful consideration of the economics of producing equivalent homogeneous mouldings for extremely long runs, unless a multi-station mould carriage is used. The use of multi-station units does pose restrictions (particularly on rotary table systems) on the size and weight of moulds acceptable, although this problem may be partially offset by the range of shot weight combinations which may be incorporated.[9]

(4) The high injection speeds necessary to fill complex mould configurations necessitates the use of high-MFI (melt flow index)

polymers, resulting in limitations on physical properties, i.e. impact properties.

### 2.2.2 Integral Skin Moulding

Integral skin moulding is a logical progression in the development of a moulding technique for the elimination of 'out-of-mould' surface finishing problems, with improved and more uniform physical properties, together with reduced polymer costs.

Clamping forces are higher than those used in conventional low-pressure moulding and the process could probably be defined as a medium clamping pressure system. For the purpose of comparison it has been included in the low-pressure moulding section to illustrate the pattern of process development (see Table 2).

The original concept of integral skin foam moulding was developed by ICI (UK) as a high-pressure moulding technique which also involved the use of a retracting mould face to induce foaming after injection.[10]

The process was subsequently, by mutual legislative agreement, licensed by Battenfeld, FRG, who, after extensive development, produced a medium-pressure integral skin foam moulding technique utilising multi-channel nozzles for injection in conjunction with melt-metering control systems.[11] Details of the relevant stages in the development of the integral skin moulding process are shown in Table 3.

*2.2.2.1 Advantages of integral skin moulding.* The obvious advantages of this moulding technique include:

(1)  the elimination of 'out-of-mould' finishing operations is possible;
(2)  the variations in the thickness of both the foam core and the homogeneous outer layers give reasonable freedom in the selection of polymers in either level layer which permits the design of mouldings to satisfy environmental and/or mechanical design considerations;
(3)  the equipment currently available has increased locking forces permitting its use, when necessary, for the production of homogeneous mouldings which utilise the total shot capacity (i.e. combined foam and homogeneous shot weights).

*2.2.2.2 Disadvantages of integral skin moulding.* The disadvantages of this moulding technique are minimal; however, various points should be noted.

TABLE 2
CELLULAR INJECTION MOULDING PROCESS

| Process name | No. of inj. stations | Type of blowing | Mould pressure | Surface condition | Mould action | Machine action | Machine type |
|---|---|---|---|---|---|---|---|
| Dow | One | Trichloromethane | Low | Poor | None | None | Mod. CIM[a] |
| UCC | One | Nitrogen | Low | Fair | None | None | Special |
| UCC-Mod. | One | Chemical | Low | Fair | None | None | Special |
| Variotherm | One | Chemical | Low | Good | Heat–Cool | None | Special |
| Dow–TAF | One | Chemical—Freon | Medium | Good | Motion req.—pressurised | None | Special |
| Allied | One | Chemical | Medium | Good | None—pressurised | Egression | Special |
| Bulgarian | One | Chemical | Medium | Good | None—pressurised | Reservoir | Special |
| USM | One | Chemical | High | Good | Motion Req. | None | Mod. CIM[a] |
| ICI | Two | Chemical | High | Excellent | Motion Req. | Sequential injection | Special |
| Hanning | Two | Chemical | High | Excellent | None | Simultaneous injection | Special |
| Battenfeld | Two | Chemical | Medium | Excellent | None | Sequential injection | Special |
| Battenfeld | One | Chemical | High | Good | Pressurised | None | Mod. CIM[a] |

[a] CIM, conventional injection moulding.

TABLE 3
INTEGRAL SKIN MOULDING TECHNIQUES

| Process | Moulding system | Nozzle configuration | Inject sequences[b] | Comments |
|---------|-----------------|----------------------|---------------------|----------|
| Original ICI system | High-pressure (retracting face) | Single channel | (a) Skin layer <br> (b) Cellular core <br> (c) Skin layer <br> Foaming induced by the reduction in melt pressure | Found limited application, but was introduced very early in the development of foam moulding |
| Battenfeld Stage 1 | Low-pressure[a] | Twin channel | (a) Skin to face opposite the gate <br> (b) Simultaneous core/skin <br> (c) Skin injection to seal the gate area | Most widely used system, can be used with suitably designed sprue and runner systems for multi-cavity systems |
| Battenfeld Stage 2 | Low-pressure[a] | Three channels | As above | Whilst developed for mouldings with long flow patterns, e.g. containers, the system is currently restricted to centre-gated single-cavity systems |

[a] Whilst referred to as 'low-pressure', clamping forces are 2·5–4·0 times greater than conventional low-pressure systems.
[b] Injection sequences are schematic with regard to Battenfeld systems since control valve systems allow variations in both feeds to permit mixed injection streams and variations in injection speeds during feeding in any channel to control skin thickness. Other factors to be considered include melt temperature and viscosity, together with mould surface temperature if surface break-through problems are to be minimised.

(1)  The design of the mould requires careful consideration with respect to the melt-flow paths, i.e. the siting of pins and inserts to minimise turbulence, which will result in mixing of the foam and homogeneous melts giving surface breakthrough problems.

(2)  The selection of polymer grades and types is critical; for instance, the adhesion between the cellular core and the homogeneous skin is essential to eliminate voids at the interface and loss of physical properties.

(3)  The shrinkage of the homogeneous outer layer resulting in distortion of the finished product after extraction from the tool necessitates the use of low-MFI polymers for the cellular core to prevent compression during the cooling cycles.

(4)  In view of the different melt fluidity of the foam and the homogeneous melt the size and siting of vents is more critical, requiring increased time allowance for mould setting and test running.

In conclusion, whilst the cost of the moulding equipment increases by comparison with the conventional low-pressure moulding technique based on the use of chemical expansion systems, the flexibility of operation resulting from the increase in clamping capacity and the elimination of 'out-of-mould' finishing operations compensate for the initial capital expenditure incurred. Comparative costings for a range of cellular moulding techniques are given in Table 7.

## 2.3 High-Pressure Moulding Systems

High-pressure moulding systems operate on the basis of maintaining an homogeneous melt phase containing a dissolved gas within the mould cavity after injection. The foaming of the polymer melt is induced by a reduction in melt pressure within the mould cavity to a level below the solution pressure of the dissolved gases. The mechanics of achieving the conditions necessary to produce cellular thermoplastic mouldings with integral skins or high-density outer surfaces form the basis of the processes described below and listed in Table 2.

### 2.3.1 Retracting Face Moulding Techniques

The reduction in melt pressure within the mould cavity to induce foaming of the polymer melt is obtained by mechanical means, i.e. the retraction of a mould face, or a retracting pin. The introduction of retracting faces and/or pins, together with the increased internal mould pressures (2000–2500 psi;

$14$–$17\,\mathrm{MN\,m^{-2}}$) necessary to ensure pressures exceed the gas solution pressure during and immediately after injection, both increase mould construction costs and restrict the configuration of moulded parts.

Initial development on this moulding technique was on rotary table equipment used in the production of unit soles, where the incorporation of a retracting base plate did not pose problems of mould design.

Mouldings exceeding $350$–$450\,\mathrm{g}$ shot weight are not known to have been produced commercially and development work has ceased in this area.

### 2.3.2 High-Pressure Integral Skin Moulding Techniques

The concept of integral skin foam mouldings, where an accurate control over both skin thickness and core density with good cell quality (particularly with the elimination of voids) resulting in mouldings having uniform reproducible physical properties was particularly attractive. Extensive development work by ICI Plastics Ltd on both the processing and equipment design resulted in a patented technology available for licence.

Whilst several options to manufacture equipment were granted, the commercial production of integral skin cellular cored mouldings using this high-pressure moulding system is thought to have ceased.

### 2.3.3 Comments on high-pressure moulding systems

In conclusion, the high-pressure foam moulding techniques, particularly the integral skin moulding system, represented the ultimate in process development for cellular mouldings. Advantages of the process include the following.

(1) High-quality component surfaces, comparable with homogeneous mouldings, are attainable. With the conventional retracting plate system, mouldings were produced at densities of $0{\cdot}3$–$0{\cdot}6\,\mathrm{g\,cm^{-3}}$ in both impact polystyrene and polyolefins, with surfaces requiring no out of mould finishing, or minimal preparation before painting. Olefin mouldings having a pleasant pearl-like surface finish were also available using this moulding system.

(2) The major advantage of both moulding techniques is the control over cell quality and density gradients through the wall of the mouldings. This control of both surface thickness and density gradients in the moulded part are directly related to mould surface temperature and the time elapsed after injection prior to the onset of pressure decay to induce foaming, i.e. the retraction of the pin(s) and/or mould face(s).

The economic disadvantages in terms of machinery/processing costs offset the performance advantages outlined above for mouldings produced using the high-pressure moulding integral skin moulding techniques. It is, however, significant that ongoing development work has in some cases successfully resulted in a number of 'hybrid' moulding techniques combining features of both high- and low-pressure moulding, in an attempt to produce mouldings with physical properties equivalent to those obtained using the high-pressure system.

## 2.4 Specialist Foam Moulding Techniques

As suggested above, a number of foam moulding techniques have been developed, some of which are of commercial importance. Details of major processes are listed below in order to complete the survey of process development.

### 2.4.1 The 'Gas Counter Pressure' Process

The principle of this process is to pressurise (e.g. at 600–800 psi; 4·0–5·3 MN m$^{-2}$) a sealed mould cavity during the filling of the tool to prevent expansion of the polymer melt on entry into the mould cavity. Immediately after injection, the gas pressure within the cavity is reduced at a predetermined rate, controlling the expansion rate and permitting the formation of a non-cooled non-cellular surface on the moulding by preventing gas 'breakthrough' on the melt front as the cavity is filled.

Practical problems restricting the extensive use of this process include those listed below.

(1) The sealing of tools, involving the use of silicone rubber seals on the closing faces (particularly tools of complex geometry) has proven problematical.

(2) The timing and sequence of mould pressurisation (counter-pressure), high-speed injection of the gas saturated melt (chemical or physical expansion systems) and the rate of pressure decay, which occur virtually instantaneously, are critical. The design of complex tools is therefore complicated and the siting and operation of extraction valves for both pressurisation and/or exhaustion can prove difficult.

(3) The polymer selection is critical and high-MFI polymers of selected molecular weight distribution are necessary for high-gloss surfaces when using styrenics.

(4) The density reduction is less than for conventional structural foams—approximately 5–8%.

The advantages of the process are again the elimination of out of mould finishing operations, with estimates of savings of 10–15% on parts costs being claimed.[12] Applications in Western Europe include marine parts and 'point-of-sale' display units where both surface quality and colour matching in a multi-component assembly are critical. Considerable interest is also apparent in North America, where mouldings are being used in similar application areas.

### 2.4.2 Foaming Technique for Thin Walled Mouldings

This process was developed in 1981–1983 in North America but until recently was little known in Western Europe. This process utilises conventional multi-nozzle low-pressure foam moulding equipment for the production of mouldings with a wall thickness of 0·100–0·125 in (2·5–3·1 mm) and a density reduction of 15–25%. Production is based on the use of ultra-high-MFI polymers (polyolefins–polypropylenes, MFI 20–35; HDPE, MFI 30–80; styrenics and polycarbonates), utilising in some cases vacuum-assisted filling to overcome problems of internal mould pressure inherent to the production of thin wall mouldings with long flow paths.[13]

Little information is available on the commercial status of the process but, whilst this development in foam moulding can be attributed to the availability of suitable polymers, it is of value on the following basis.

(1)  The structural foam moulding has been restricted to large shot weights (in excess of 1–2 kg) with an average wall thickness of 0·25 in (6·2 mm).

(2)  The entry of foam mouldings into the large-volume thin wall moulding market opens new markets currently dominated by homogeneous mouldings.

(3)  The surface quality, whilst anticipated to be good as a result of the use of high-MFI polymers, should require minimal finishing operations to produce cellular mouldings which are competitive with those currently supplied in the homogeneous form for non-domestic applications.

### 2.4.3 The 'TCM' (Moulding) Process

This process was introduced into Western Europe from the USA in 1983, where it has been successfully operated for several years. The 'TCM' process differs in principle from the established foam moulding techniques, utilising multi-station heated tools in combination with slow injection speeds using a two-stage extruder/plunger design. The combination of these two processing characteristics is claimed to minimise melt turbulence during injection to give mouldings with smooth surfaces which require minimal out-of-mould finishing.[14]

Little processing data or information on the selection of polymer grades is yet available for this process, which is thought to be operated by three companies in Western Europe.

### 2.4.4 Crosslinked Polyolefin Moulding

The production of cellular crosslinked ethylene vinyl acetate materials is not novel; such materials are currently produced in sheet form using the compression moulding techniques as described in Section 4. An interesting development is the production of such materials using injection-moulding techniques based on the simultaneous crosslinking and expansion of ethylene vinyl acetate copolymers in a heated mould cavity.[15,16] The use of the moulding technique permits the production of relatively complex shapes, minimises secondary fabrication processes, and reduces problems of scrap utilisation inherent to press moulding. Important characteristics of the moulding technique include the following.

*2.4.4.1 Moulding equipment.* Mouldings are produced using conventional 'in line screw' moulding equipment, but this differs from conventional foam moulding techniques in not requiring a shut-off nozzle, or high injection speeds.

Screw design is important, together with an accurate control over melt temperature during processing to maximise mixing by the screw. Both the crosslinking system, an organic peroxide (dicumyl peroxide) and the expansion system, normally a highly activated grade of azodicarbonamide, must be fully dispersed by the screw without thermal activation. In practice the melt temperature during mixing does not exceed 100°C and the screw dwell time is minimised by delaying commencement of the plasticisation cycle.

*2.4.4.2 Mould design.* Heat transfer into the mould cavity can pose problems and a uniformity of heat transfer over the total mould surface is essential. Particular attention is necessary in:

(1)   the insulation of the tool in the area of the parting line, where small decreases in surface temperature reduce the level of crosslinking in

the polymer melt, resulting in the splitting of the outer skin during
ejection of component from the mould cavity;

(2) the large instantaneous volume expansion of the moulding
(100–150% by volume) which occurs on opening of the mould
requires attention to both mould wall 'draw angles' and possible
'undercuts', which can result in tearing and splitting at the mould
ejection/expansion stage;

(3) whilst total insulation of the hot oil, or electrically heated tool, is
necessary, the use of an insulated sprue bush is essential (heat
transfer from the injector nozzle will result in localised heating in the
gate area causing premature activation of the organic peroxide
leading to nozzle blockage and irregular filling of the mould cavity);

(4) that normal mould clamping forces are used to compensate for high
internal mould pressures resulting from the high levels of the
expansion system used;

(5) the shrinkage of the ejected mouldings during cooling may
necessitate the use of internal shaping jigs for certain mouldings, e.g.
tyre production.

*2.4.4.3 Polymer selection and formulation.* A novel aspect of the process
is the ability to formulate compounds to achieve specific performance
characteristics, e.g. density reduction, hardness, resilience, or abrasion
resistance. by the choice of fillers, polymer systems, etc., analogously to
work carried out in the area of press moulding. It must however be
emphasised that the process can be extremely sensitive to minor variations
in both compounding and processing variables.

Major considerations in the selection of polymers include:

(1) *high melt fluidity* in the temperature range 100–120°C is necessary to
permit the compounding of both the crosslinking and expansion
systems without pre-decomposition and subsequently to ensure
good mixing of the compounds, without decomposition on a
25–30:1 $L/D$ screw during the moulding operation;

(2) *rates of crosslinking* are critical in attaining the necessary balance
between melt elasticity and cell formation. Low levels of
crosslinking resulting from a low-reactivity polymer, or the
incorrect selection of peroxides, result in skin rupture and splitting
of the moulding during expansion. High levels of crosslinking give
low expansion and high levels of shrinkage during cooling after
expansion/ejection from the mould cavity, attributable to the low
elasticity, or 'memory', of the outer surface of the moulding during
cooling.

The above polymer processing characteristics have resulted in the use of ethylene vinyl acetate copolymers, MFI 2·0–4·5, with vinyl acetate levels of 9–14·0% by weight. Considerably more work on the chracterisation of suitable olefin systems for use in the process is necessary. Practical tests have shown that copolymers of a similar vinyl acetate content, but produced by differing polymerisation routes, show differences in both melt strength and melt elasticity after crosslinking. The differences are assumedly attributed to the differences in molecular weight distribution and the frequency and length of the side chain branching in the polymers.

Other polymer blends which are being investigated include EVA/synthetic rubber, EVA/acrylate copolymerisation additives, and filled compounds containing silane-type flow promoters in order to improve the surface melt strength and to reduce the effects of thermal processing variables.

The areas of application for these mouldings include flotation applications, e.g. boat fenders, fishing floats (only suitable for use in shallow water, maximum depth 30 ft (9 m)), tyres for toys, golf trolleys, industrial trolleys and unit sole production in the shoe industries.

The physical properties for these materials are very similar to those given for press-moulded materials as shown in Table 9.

## 2.5 Discussion

A comparison of Table 1 and 2 illustrates the relationship between the process development and increased performance requirements for moulded parts supplied to the major application areas, e.g. computer, business machine housing and audio/visual markets. Whilst the pattern in process development is logical and could be anticipated, the installation of advanced foam moulding equipment remains slow. Differences in the utilisation of advanced foam moulding techniques between Western Europe and North America are apparent, with low-pressure moulding techniques dominating the North American market. Reasons for the differences in the pattern of process development may include the wider use of direct gas injection foam moulding techniques and the lack of furnishing applications (excluding upholstered frame and reproduction applications) in the North American market.

However, the effects of the economic recession have resulted in a more critical assessment of the economics of the structural foam moulding process, particularly low-pressure moulding techniques where 'out-of-mould' finishing costs constitute a major proportion of the total part production costs. It is felt that a direct result of this analysis regarding the operating economics, together with the competition from both homogeneous mouldings alone, or in combination with cellular thermo-plastics and competitive processes, (i.e. urethane/nylon Reaction Injection Moulding (RIM) processes in the major markets) will be a more rapid acceptance of the advanced foam moulding techniques.

## 2.6 Developments in Polymer Systems

For convenience, developments in each major class of polymer have been summarised individually below.

### 2.6.1 Polyolefin Resins

A disadvantage of cellular mouldings produced from polyolefin resins is the characteristic 'wax' surface and the need for additional treatment if surface finishing is necessary. Applications for polyolefin mouldings have been restricted to glass-reinforced and talc-filled materials, for use in industrial applications, e.g. automotive or domestic appliance parts where smooth unfinished surfaces and water resistance, particularly of polypro-pylene, are an advantage. Domestic applications are limited but applications in furniture construction are known in Western Europe.[17]

Development of polyolefin grades specifically used for foam moulding is mainly restricted to 'coupled glass' filled grades of polypropylene homo- and co-polymers, where problems of surface staining have occurred when using conventional glass fibre-filled compounds. The problem is attributed to migration of the fibre coupling agent during foaming. Developments in thin wall foam moulding in North America, as discussed above, have also resulted in the introduction of ultra-high-MFI grades of olefins in the USA, but such resins are not thought to be available in Western Europe in commercial quantities.

### 2.6.2 Styrenics (Including Copolymers and Alloys)

Development in the resins in this area has been stimulated by the increasing legislation on flammability, improved surface smoothness requirements (particularly in low-pressure moulding techniques) and the need to compete with engineering thermoplastics in certain application areas, i.e. computer and business machine areas.

Developments in styrenic homopolymers, copolymers, terpolymers and alloys are outlined below.

*2.6.2.1 Styrene homopolymers.* The introduction of the ultra-high-impact grades of polystyrene which also satisfy the minimal V-O (ASTM-UL954) performance requirements, has increased their penetration of the business-machine markets for structural foam mouldings. A number of grades are available which are suitable for use with both chemical and physical expansion systems. Such grades of polystyrene have excellent flow properties resulting in smooth surfaces and short cycle times and do not give problems of vent blockage, or deposits on the mould surfaces.

*2.6.2.2 Copolymers and terpolymers containing styrene.* Little activity is apparent in this area, but some developments are apparent in flame retardant grades of ABS containing pre-compounded inorganic expansion systems. Suitable grades of ABS are available in both Western Europe and North America, but their use is not widespread as a result of poor economics based on a performance (impact properties, cycle times)/cost basis by comparison with copolymers, or styrenic alloys.

*2.6.2.3 Styrenic alloys.* The alloys can be classed as semi-engineering thermoplastics since the physical properties of these materials when foamed are superior to both styrene homopolymers and terpolymers. Developments of styrenic alloys have taken place using polyphenylene oxide and polycarbonate as the alloys.

(a)  *Polyphenylene oxide/styrene alloys* constitute the most widely used polymer system for the production of video and business machine housings. This group of alloys are the subject of an ongoing development programme intended to:

  (1)  improve surface properties and flow characteristics particularly for use in gas counter-pressure moulding techniques;
  (2)  improve flame-retardant characteristics without the loss of impact properties and surface finish;
  (3)  improve heat and solvent resistance by the blending of PPO with nylon;
  (4)  improve UV stability, particularly in office equipment applications where prolonged exposure to artificial light occurs.

  This has resulted in the commercial availability of a range of grades suitable for foam moulding which possess improved moulding and performance characteristics.

(b)  *Polycarbonate/ABS alloys* are a recently developed range of materials suitable for the production of cellular mouldings with good physical properties and flame resistant characteristics. The component parts of the alloys are complementary in:

  (1)  improving the thermal stability of polycarbonate alone;
  (2)  improving the surface smoothness of both polymers;

(3)   enhancing the impact/environmental performance characteristics.

A disadvantage remaining is, however, the degradation and staining of the polycarbonate and ABS components respectively when ammonia-liberating chemical expansion systems are used. Little information is yet available on these materials, but published articles indicate their use in cellular moulding in both the USA and Japan.

### 2.6.3 Engineering Thermoplastics

Whilst the engineering thermoplastic polymers have received considerable attention, in practical terms only polycarbonate, the thermoplastic polyesters (polyethylene terephthalate and polybutylene terephthalate) and the polyamides have found limited commercial applications.

The loss of physical properties (particularly impact strength) and the surface quality, has restricted their use to speciality moulding applications. It is thought that this situation could change with the increased use of the integral skin foam moulding techniques. However, the resultant mouldings could be composite structures utilising the high-performance polymers in the outer skins and cheaper lower-performance polymers for the cellular core.

### 2.6.4 Discussion

As indicated above, work is continuing on the development of polymers intended to improve both the reproducibility of physical properties and the quality of surface finish in conventional low-pressure moulding techniques. In addition, considerable activity is apparent[18] in the development of polymer alloys for use in counter-pressure moulding operations where the effects of the internal mould gas pressure is to reduce melt flow path lengths, and conversely in the development of polymer systems of limited molecular weight distribution to satisfy the requirements of both the thin wall and integral skin foam moulding processes.

## 2.7 Developments in Expansion Systems

For convenience, developments in expansion systems specifically for the use in cellular injection moulding can be divided into two areas.

### 2.7.1 Physical Expansion Systems

Developments in the use of direct nitrogen injection techniques remain static, with no indication of investigations into the use of gaseous mixtures or nucleation systems to improve cell and/or surface quality of the resultant mouldings. Developments in the use of physical expansion systems based on the use of pentane or fluorocarbon mixtures in the moulding of polystyrene[19] have been reported. Improved surface quality, together with low densities and fine cell structures, are claimed. These properties can be attributed to the plasticisation of the styrene melt to give enhanced flow and melt elasticity at low temperatures, reducing cell coalescence to a minimum. Since the composition of the gaseous expansion system is of major importance in obtaining uniform cell structures and hence surface smoothness, it is thought that investigations involving the combination of both physical and chemical expansion systems constitute a logical future development area.

### 2.7.2 Chemical Expansion Systems

Again for convenience, developments can be divided into two distinct areas:

*2.7.2.1 Inorganic expansion systems.* Recently considerable interest has arisen in the use of expansion systems which liberate carbon dioxide by the interaction of an inorganic carbonate and a weak acid (i.e. citric acid or sodium citrate). Advantages claimed include:

(1)  improved cell quality and reduced cycle time, probably attributable to the endothermic nature of the acid/base reaction and/or the cooling of the polymer melt by the high-melt-permeability characteristics of the carbon dioxide;
(2)  they are suitable for use in ammonia-sensitive resins and when using copper electroform tools;
(3)  they can be used in 'food contact' applications and have FDA and BGA clearance, leaving no residual odour in the mouldings.

Polymer systems incorporating acid/carbonate expansion systems are available based on ABS, but the initial expandable styrene moulding compounds were withdrawn in Western Europe. Whilst reasons for the withdrawal of such compounds vary, e.g. moisture pick-up and variation in gas yield, this type of expansion system in a masterbatch or encapsulated form will find application in specific cellular moulding applications.[19]

The use of metal hydrides as a source of hydrogen in the expansion of polystyrene mouldings has been reported in the USA[20] but it is not thought to be of commercial importance at present.

*2.7.2.2 Organic expansion systems.* Whilst numerous patents have been published suggesting novel expansion systems, commercial development has concentrated on two classes of chemical expansion system.

(1)  *Modified grades of azodicarbonamide:* As a result of their high gas
     yield, ease of activation and, of major importance, the low toxicity
     of both the evolved gases and residual decomposition products,
     azodicarbonamide blends have been extensively developed for use in
     cellular injection moulding. Stages in the development of blends of
     azodicarbonamide for use in injection moulding are shown in
     Table 4.

     Whilst the azodicarbonamide blends listed comprise simple
     physical blends, the function of the additive may vary considerably.
     The major function of the additive in most cases will be to reduce the
     decomposition temperature of the azodicarbonamide, but a
     secondary, possibly more important, function can include the
     modification of the composition regarding the evolved gas and/or
     the temperature range over which gas is evolved, resulting in
     changes in gas permeability to improve cell quality and surface
     smoothness.

     A comparison of the gas permeability rates is shown in Table 5.
     Examination indicates the similarity between the expansion
     characteristics of the gas mixture normally evolved from organic
     chemical systems and the fluorocarbon mixtures used in the
     expansion of low-density polystyrene. The major difference is the
     lack of melt plasticisation by the gases listed in Table 5. As a result,
     the sole function of the gas is to stabilise the cell structure during
     cooling to eliminate shrinkage. High permeability rates are
     therefore not desirable—hence the predominance of nitrogen as the
     major component of the gaseous mixture evolved on decompo-
     sition. This is true for the majority of commercially available
     expansion systems.

     In addition, modifications are continuing to be made to the
     decomposition mechanism of azodicarbonamide which eliminate
     gaseous or solid decomposition residues which react with cure
     accelerators or give coloured residues.

     Development work is therefore proceeding on both the physical
     and chemical modification of azodicarbonamide and further
     modified grades can be anticipated in the near future.

(2)  *High-temperature blowing agents (decomposition point above
     220°C):* This area is currently the subject of several research
     programmes. Development is divided into two distinct areas,
     compounds which on decomposition evolve ammonia as part of the
     gaseous decomposition products and those which do not. Main

TABLE 4
COMPOSITION OF COMMERCIALLY AVAILABLE AZODICARBONAMIDE BLENDS

| Type | Typical composition | Decomposition range (°C) | Areas of end application |
|---|---|---|---|
| Conventional metal salt activation | Azodicarbonamide/ zinc oxide | 165 | Injection moulding of polystyrene, LDPE, EVA and thermoplastic rubbers |
| Addition of a secondary chemical blowing agent, acting as an activator and an additional gas source | (a) Azodicarbonamide/ dinitrosopentamethylenetetramine (DNPT) | 150 | Injection moulding of EVA, thermoplastic rubbers, rubber polyolefin blends and crosslinked systems |
| | (b) Azodicarbonamide/urea | 197 | |
| | (c) Azodicarbonamide/ benzenesulphonhydrazide | 161 | |
| Combination of both above systems | (a) Azodicarbonamide/ silica/zinc oxide | 190 | Injection moulding of styrenes, polyolefins, crosslinked olefins |
| | (b) Azodicarbonamide/urea/ micro-crystalline wax/ dinitrosopentamethylenetetramine (DNPT) | | Finding wide application in rubber/polyolefin mixtures, crystalline wax aiding dispersion in 'rubber-like' stocks |

TABLE 5

PERMEABILITY OF THERMOPLASTIC FILMS TO GASES

| Film material | Permeability of film[a] | | | |
|---|---|---|---|---|
| | Nitrogen | Oxygen | Carbon dioxide | Water vapour |
| Polyvinylidene chloride | 0·009 4 [1] | 0·053 [6] | 0·29 [31] | 14 [1 490] |
| Polychlorotrifluoro- ethylene | 0·03 [1] | 0·10 [3] | 0·72 [24] | 2·9 [97] |
| Polyethylene terephthalate | 0·05 [1] | 0·22 [4] | 1·53 [31] | 1 300 [26 000] |
| Polyamide (nylon 6) | 0·20 [1] | 0·38 [4] | 1·6 [16] | 7 000 [70 000] |
| Unplasticised PVC | 0·40 [1] | 1·20 [3] | 10 [25] | 1 560 [3 900] |
| Cellulose acetate | 2·8 [1] | 7·8 [3] | 68 [24] | 75 000 [26 800] |
| Polythene (0·954–0·960 density) | 2·7 [1] | 10·6 [4] | 35 [13] | 130 [48] |
| Polythene (0·122 density) | 19 [1] | 55 [3] | 352 [19] | 800 [42] |
| Polystyrene | 2·9 [1] | 11 [4] | 88 [30] | 12 000 [4 100] |
| Polypropylene | — | 23 [4] | 92 [16] | 680 [118] |
| Ethyl cellulose (plasticised) | 84 [1] | 265 [3] | 2 000 [24] | 130 000 [1 500] |

Data source: Briscall, H. and Thomas, C. R., Cellular structure and physical properties of plastics. *Brit. Plast.* (July 1968), 79–84.

[a] The ratios of the permeabilities were calculated for each film and are shown in square brackets after the actual permeabilities in $cm^3/cm^2/mm/s/cm$ Hg $\times 10^{11}$.

interest is in the latter group of compounds as a result of the increased potential applications for cellular engineering thermoplastics, i.e. polycarbonate, polyethylene terephthalate and polyethersulphones. Details of products reported to be commercially available are given in Table 6 (this list does not include numerous experimental compounds known to be available).

## 2.7.3 Speciality Expansion Systems

A recent, interesting area of development covering all the above types of chemical expansion systems is the increasing use of both melt concentrates and liquid dispersions. Preliminary examination of the basis for these developments indicates a reduction in the handling of fine powders, reducing environmental risks to operatives. Additional benefits have been noted, e.g. increased gas retention due to the carrier system acting as a cell nucleator. Other advantages are also possible, e.g. a multi-functional

TABLE 6

HIGH-TEMPERATURE EXPANSION SYSTEMS

| Type | Chemical composition | Operating temperature range (°C) | Gas yield (cm³ g⁻¹) | Comment |
|------|---------------------|-----------------------------------|---------------------|---------|
| Celogen[a] CB | Patented hydrazine | 170–220 | 200 | No ammonia evolved |
| Celogen[a] BH | 4,4'-Oxybis(benzenesulphonyl semicarbazide) | 210–220 | 154 | No ammonia evolved |
| Celogen[a] RA | p-Toluene sulphonylsemicarbazide | 220–245 | 140 | |
| Celogen[a] HT 550 | Patented hydrazide | 250–310 | 155 | No ammonia evolved |
| Genitron[b] AF100 | Patented composition | 220–250 | 125 | No ammonia evolved |
| Expandex[c] 5PT | 5-Phenyltetrazole | 250–280 | 200 | Low ammonia content |
| Kemtec[d] 300 | Patented compounds stated to be 'neutral aromatic organics' | 155–230 | 157 | |
| Kemtec[d] 500 | | 290–320 | NA | |
| Kemtec[d] 550 | | 290–320 | NA | No ammonia evolved |

NA, data not available.
[a] Trade mark of Uniroyal Chemical Co., USA.
[b] Trade mark of Schering Industrial Chemicals.
[c] Trade mark of Olin Chemicals, USA.
[d] Trade mark of PMC Specialistics Group Inc., USA.

carrier system acting as a combined decomposition activator, source of a crosslinking activator and/or free-radical source.

More advanced developments in chemical expansion systems include the investigation into the possibility of initiating decomposition by the use of high-frequency heating techniques.

These developments, independent of the type of expansion system incorporated into the carrier system, result in the automation of the cellular moulding operation on lines similar to those currently apparent in the moulding of homogeneous parts, thus improving the economics of the cellular moulding processes.

### 2.7.4 Discussion

As summarised above, considerable changes in the type of chemical compound available and the form in which it is available are occurring and it is anticipated that these developments will continue. It is also apparent that some overlap between physical and chemical expansion technologies could occur if the production of cellular mouldings with low densities, fine cell structures and surfaces which require minimal 'out-of-mould' finishing is to be achieved, particularly in the 'low-pressure' moulding area.

## 2.8 An Economic Comparison of the Major Foam Moulding Techniques

Information on the comparative economics of the various structural foam moulding techniques is, of necessity, mainly limited to theoretical studies carried out by major polymer producers, or manufacturers of foam moulding equipment. A paper presented at the *Fifth Structural Foam Conference* in May 1977[21] compared the costs of parts produced using the conventional low-pressure moulding technique, with the cost of producing similar parts using thermocycling, gas counter-pressure and low-pressure integral skin foam moulding techniques. Details of the comparative total manufacturing costs are shown in Table 7 and it is felt that the differences shown will remain very similar in 1988–1989.

### 2.8.1 Cost Comparison of Typical Processes

The details of mouldings used as basis for comparisons are as follows.

*Product A: Toilet water tank (high-impact polystyrene)*
Surface: high gloss
Parts/year: 400 000
Mould cavities: four
Steel tool

*Product B: Business machine component (polyphenylene oxide/styrene alloy)*
Surface: textured, but will be painted to colour-match homogeneous plastic and metal parts
Parts/year: 10 000 over five-year period
Mould cavities: two
Aluminium (Al) tool for low-pressure moulding
Steel mould for other processes

*Product C: Bucket seat (talc-filled polypropylene)*
Surface: textured (engraved mould surface)
Parts/year: 150 000 over three-year period
Mould cavity: one
Steel tool

TABLE 7

ECONOMIC COMPARISON OF THE MAJOR FOAM MOULDING TECHNIQUES

| Product | Process | Material cost (%) | Moulding cost (%) | Mould cost/part (%) | Finishing cost (%) | Total Manuf. cost (%) |
|---|---|---|---|---|---|---|
| (A) Toilet water tank (high-impact polystyrene) | SF | 100 | 100 | 100 | 100 | 100 |
| | Thermocycle | 100 | 169 | 125 | 0 | 19 |
| | Counter-pressure | 125 | 123 | 100 | 39 | 53 |
| | Int. skin | 164[a] | 166 | 125 | 0 | 28 |
| (B) Business machine component PPO/styrene (Noryl) | SF | 100 | 100 | 100 | 100 | 100 |
| | Thermocycle | 100 | 171 | 111 | 40 | 84 |
| | Counter-pressure | 117 | 131 | 136 | 40 | 92 |
| | Int. skin | 83[b] | 169 | 132 | 40 | 82 |
| (C) Bucket seat (talc-filled PP) | SF | 100 | 100 | 100 | 100 | 100 |
| | Thermocycle | 100 | 174 | 107 | 57 | 87 |
| | Counter-pressure | 111 | 136 | 104 | 57 | 87 |
| | Int. skin | 127[c] | 144 | 100 | 0 | 65 |

[a] Material cost involves ABS skins, ABS re-grind foam core.
[b] Material cost involves Noryl skins, polystyrene foam core.
[c] Material cost involves PP skin, talc-filled PP foam core.

For ease of comparison, the costs for low-pressure foam moulded parts are equal to 100%. Cost savings and increased costs of comparative moulding techniques are calculated as a percentage of the low-pressure foam moulding process cost.

*2.8.1.1 Breakdown of costs.* Costings for the individual processing techniques were derived as follows for the comparison with the conventional low-pressure moulding technique.

(a)   *The thermocycling technique*

    (1)   *Material costs* should be similar to low-pressure moulding.

    (2)   *Moulding costs* increase due to increased cycle times and the additional cost of ancillary thermal cycling unit.

    (3)   *Mould costs* increase due to complexity of mould design.

    (4)   *Finishing costs* assume single-coat paint applied.

(b)   *The counter-pressure moulding technique*

    (1)   *Material costs* increase due to minimal density reductions (5–10% reduction) in addition to higher-cost speciality high-gloss resins.

    (2)   *Moulding costs* increase due to slight increase in cycle time in addition to ancillary equipment costs. Equipment costs will also increase due to the necessity for increased clamping forces.

    (3)   *Mould costs* increase slightly dependent on the level of internal pressure required, i.e. for aluminium, or steel tool.

    (4)   *Finishing costs*—since streaking can occur, provision for a single coat of paint has been made with the resultant slight increase in total production costs.

(c)   *The low-pressure integral skin moulding technique*

    (1)   *Material costs* increase due to increase in overall density, i.e. homogeneous skin, but may be offset by the use of regrind or lower-cost resin in the cellular core.

    (2)   *Moulding costs* increase due to higher cost of moulding equipment, e.g. two injection units, higher clamping forces.

    (3)   *Mould costs* are slightly increased by improved surface quality requirements.

    (4)   *Finishing costs* are virtually eliminated.

Whilst the comparative costs shown in Table 7 indicate the benefits of the low-pressure integral skin moulding techniques, the selection of moulding technique will depend on the performance requirements of the mouldings, particularly surface quality. Finishing costs (estimated 40–50% of total part cost) still remain the major factor in determining the economic production and acceptance of a cellular moulding in preference to homogeneous moulding.

## 2.9 Conclusions

Whilst the rate of growth in structural foam moulding has slowed considerably in comparison with the growth observed during the peak period 1974–1979, it is predicted that the level of polymer consumption during 1988–1989 should reach similar levels to 1974–1979. This prediction can be substantiated by:

(1) the increasing number of integral skin moulding machines being installed, particularly in the USA, where the continuing use of physical expansion systems has slowed down the acceptance of structural foam parts in semi-metal replacement applications;

(2) the increased range of resin types and relevant expansion systems available;

(3) the increasing volume of physical property data now being generated on cellular materials, making the resultant mouldings of more interest to designers and engineers.

In the writer's opinion the increased growth will not result directly from any single factor suggested above, but a combination of all three factors should lead to an increased acceptance of structural foam mouldings alone, or in combination with homogeneous mouldings in a wider range of end application areas.

## 3 THE EXTRUSION OF CELLULAR THERMOPLASTICS

The area of extruded cellular thermoplastics is not as widely developed as cellular injection moulding, but it constitutes a large potential area for development, particularly with low-density crosslinked polyolefins and/or polyolefin/rubber blends and rigid profiles based on PVC or polystyrene.

### 3.1 Process Developments

The techniques used for extruding cellular products are as follows.

#### 3.1.1 Free Expansion Techniques

These techniques involve the expansion of the polymer melt immediately after leaving the main die prior to entry into the vacuum/shaping dies, and remain the most widely used processes. Despite considerable work on die design, extrudate cross-sections are restricted to $0.5–1.5\,in^2$ ($3.2–9.7\,cm^2$) due to the problems of uniformity of melt pressure across the die aperture. Development in this area remains static with the exception of complex profiles where complex die designs involve the use of restricting grids and/or multi-centre mandrel systems in an attempt to minimise melt pressure variations across the extruded profile.[22]

*3.1.2 Controlled Expansion Techniques.* These are based on the Celuka process and include the use of a tapering mandrel projecting through the main die into the shaping die. This die configuration results in a control over the pressure decay in the polymer melt (i.e. rate of expansion) and eliminates surface problems encountered in free expansion during the production of complex profiles. The die and shaping unit design developments are restricted to licensees of the patented Celuka process.[23]

*3.1.3 Co-extrusion Techniques*
These are essentially a combination of the two processes listed above. The thickness and melt elasticity of the outer skin control the rate of expansion of the cellular core [i.e. it acts as an *in-situ* shaping die, permitting the production of extrudates with a low-density core ($0.2–0.3\,g\,cm^{-3}$) with homogeneous outer skins]. This area is the subject of an on-going development by machinery/polymer producers and is not thought to be the subject of patent restrictions.[24]

*3.1.4 Crosslinked Extrusion Techniques*
These techniques involve secondary downstream curing equipment based upon the use of hot air, salt baths or high-frequency heating in order to achieve sequential crosslinking and expansion of the extrudate.[25–27] Whilst it is accepted that the extruder is basically a mixing and shaping unit for the unexpanded/non-crosslinked sheet or profile, it appears logical to include the process under extrusion technology. This area of technology is the subject of considerable development in polymers, expansion systems, crosslinking additives and mechanical development, e.g. improved compounding/mixing screw designs with high output rates.

## 3.2 Developments in Polymers

*3.2.1 Developments in Olefin Polymers and Copolymers*
Activity in this area is on the development of suitable grades of low-density polyethylene (LDPE) for use in crosslinked processes based on chemical and/or irradiation crosslinking techniques. An additional area of

development is in cable insulation (telephone transmission cable) where line speeds on a crosshead extrusion die of 3000 ft min$^{-1}$ (915 m min$^{-1}$) necessitate low head pressure to minimise conductor breakage when using either polypropylene or medium-molecular-weight polyethylene. The basic characteristics of each process have been identified and suitable ranges of polymer developed.

*3.2.1.1 Crosslinkable LDPE.* Crosslinkable grades of LDPE homopolymers and/or vinyl acetate copolymers possessing the following characteristics are available.

(1)   a density range of 0·1 g cm$^{-3}$ and a melt flow index of 4 to 7;
(2)   highly branched structures, but with a high frequency of short chain branching (the incidence of long side chains is thought to retard crosslinking rates);[28,29]
(3)   low gel content;
(4)   in many cases unstabilised, so as to reduce the retardation rate of crosslinking by radical abstraction by the antioxidants normally incorporated.

*3.2.1.2 Cable insulation grades.* Cable insulation grades are the subject of an on-going development on the use of medium-density polyethylene (MDPE) in Western Europe and on high-density polyethylene (HDPE) and polypropylene (PP) in North America. Considerable published information[30] is available on the selection and processing characteristics of LDPE and HDPE polymers that are suitable for use in both co-extruded cellular cable, or single-layer cellular cable insulation.

*3.2.2 Styrene Developments*
Work on the development of improved expandable grades of polystyrene is mainly concentrated in North America where large markets in 'free expansion' and co-extruded domestic profiles exist. In Western Europe expandable grades (containing the necessary expansion system) are available from major producers of polystyrene, but no other developments are apparent.

*3.2.3 Developments in PVC Polymers and Blends*
The types of PVC polymers and blends used for cellular extrudates are divided into two categories.

*3.2.3.1 Rigid compounds.* A range of unplasticised expandable powder blends (with and without the addition of expansion systems) are available from major polymer producers and/or custom blenders. Formulations are based on a $K$ value of 70 for PVC homopolymers in Western Europe, with developments on both polyolefin and acetate copolymers apparent in North America.

The selection of impact modifiers is critical in the formulation of rigid compounds to be used in free expansion systems. Acrylic-type modifiers are preferred acting as melt modifiers, increasing melt elasticity to give extrudates with fine uniform cell structures and smooth outer surfaces. The correct lubrication balance is essential, in order to:

(1) ensure a homogeneous melt at the decomposition temperature of the expansion system;
(2) ensure adequate mixing without delaying gelation.

External lubrication levels will depend upon the extrusion technique used; conventional loadings of oxidised polyethylene or microcrystalline waxes are suitable for free expansion techniques. Controlled expansion techniques, e.g. the 'Celuka process', require high lubricant loadings to ensure good skin formation as the melt pressure decays within the shaping unit.

The thermal stabilisers used in expandable compounds have a dual function:

(1) to ensure the thermal stability of the compound during both the compounding and processing cycles; also
(2) to function as an activator if azodicarbonamide is being used as the expansion system.

Considerable amounts of published data are available detailing the interactions between azodicarbonamide and the conventional metal stabilisers, illustrating their use in reducing the decomposition temperature of azodicarbonamide in PVC formulations.[31,32]

*3.2.3.2 Flexible cellular compounds and polymer blends.* Developments on flexible extrudates, based on 'free expansion' techniques and utilising conventional plasticisers, are limited to the production of profiles of low cross-section. A major problem is the development of formulations to create the necessary head pressure within the die head so as to prevent problems of pre-foaming occurring. As a result, interest exists in the production of co-extruded, cellular-cored extrudates, to give improved skin quality.

The problems of die head pressure can be eliminated if the function of the extruder in this process is to operate as a mixing unit and to produce a 'pre-form' for cure and expansion as a separate stage, or on a continuous or batch production basis.

A method for the production of sheet and tube in the density range $2\cdot5$–$15\cdot0\,\text{lb}\,\text{ft}^{-3}$ (40–240 kg m$^{-3}$) has been developed based on the use of butadiene–acrylonitrile rubber/PVC blends. The use of sulphur curing elastomers as a substitute for conventional plasticiser systems permits the production of highly resilient foams with improved physical properties. However, the degree of improvement is dependent upon the choice of elastomer.

The formulation of suitable compounds will be dependent on the following factors.

(1) *The expansion technique.* Both hot-air ovens and high-frequency heating units are used for the simultaneous curing and expansion of the extrudates. The higher heating rates achieved in high-frequency heating units permit the use of ultra-fast cure systems. However, formulation developments must compensate for heating rates since a critical balance exists between the degree of crosslinking (i.e. cure rate) and the rate of expansion if low-density, fine uniform cell structure extrudates are to be produced.

(2) *The type of elastomer.* Whilst most formulations are based on acrylonitrile rubber, both polychloroprene and ethylene–propylene rubbers can be used to obtain specific physical properties, e.g. oil resistance, high compression set characteristics, or ozone resistance.

(3) *The expansion system.* As stated above, the critical balance between crosslink density and expansion rate necessitates careful examination of both the expansion system and ingredients (e.g. PVC thermal stabilisers and cure accelerators) since interactions can affect the cure rate and the temperature, and the rate of decomposition of the chemical blowing agent.

Whilst a range of chemical blowing agents have been used in this process, e.g. dinitrosopentamethylenetetramine (DNPT) and BL353, the nitroso compounds are not now used and most formulations utilise azodicarbon-amide-based mixtures. A typical mixture would be azodicarbonamide/oxybis(benzenesulphonhydrazide), where the presence of the sulphon-hydrazide acts in two roles:

(1) a method of reducing the decomposition temperature of the azodicarbonamide to 160–170°C, which is within the processing range of the compounds;

(2) the decomposition residues from the sulphonhydrazide act as synergistic cure accelerators, offsetting the cure retardation observed when azodicarbonamide alone is used in the expansion of other synthetic rubber formulations.

Obviously considerable expertise has been amassed by producers of NBR/ PVC foams on both formulations, and the processing conditions necessary to achieve the critical balance between crosslinking and expansion, but little information has yet been published.[33] Recently two published articles[33,34] illustrated the use of a modified Brabender Plasti-Corder to study the relationship between cure and expansion in both continuous and batch curing techniques. Details of the relevant rheology charts illustrating the formulation variables discussed above are shown in Fig. 1.

The differences in both the cure rate and the expansion rate for both press cure systems and continuous production techniques are clearly illustrated and further indicate the balance between the two functions when producing low-density foam materials.

Examination of Fig. 1 also indicates that the expansion/crosslinking balance in PVC/nitrile foams is very similar to that observed in crosslinked polyolefin foams discussed in Section 3.2.1.

### 3.3 Developments in Expansion Systems

As can be anticipated, most expansion systems developed for use in cellular moulding are suitable for use in cellular extrusion processing, although modified grades of azodicarbonamide have been developed specifically for use in crosslinked polyolefin foams.

For convenience, developments can be divided into two areas.

#### 3.3.1 Physical Expansion Systems

Two areas of application for direct gas injection currently exist:

(1) *cable insulation*, telephone transmission cable insulation based on gas injection but utilising small quantities of azodicarbonamide as a 'gas nucleator' is in commercial production in both Japan and the USA;[35]

(2) *sheet and profiles* based on polyolefins using the injection of gaseous mixtures are also in commercial production in Western Europe and North America.

Those applications together with the well established fluorocarbon/ polystyrene paper process constitute the major area of production of extrudates based on physical expansion systems.

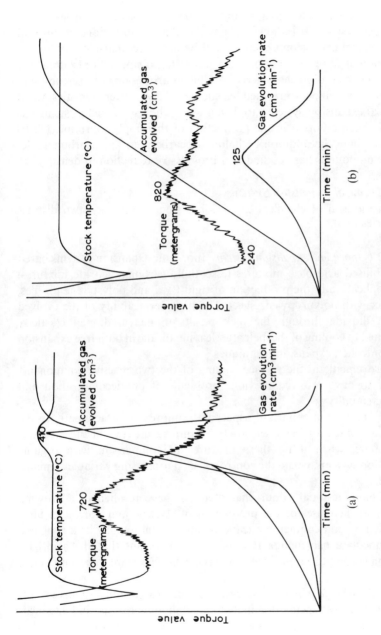

FIG. 1. Typical melt rheology/gas evolution curves for cellular NBR/PVC foam. Brabender Plasti-Corder traces for (a) batch and (b) continuous processes. Data source: Ref. 26, p. 152.

Development in this area is limited, due to problems of nucleation of single-gas systems. It is, however, possible that combinations of physical and chemical expansion systems could be developed particularly for the expansion of styrene homopolymers and/or terpolymers. For example the fugitive plasticisation characteristic of fluorocarbons on polystyrene could improve low-temperature melt elasticity to give lower densities when chemical expansion systems are being used. The use of such systems has recently been reported in the production of thermoplastic rubber (TPR) profiles where the addition of halohydrocarbons to azodicarbonamide-based compounds has resulted in a progressive reduction in density.

### 3.3.2 Chemical Expansion Systems
Within the field of chemical expansion system technology, two distinct categories exist.

#### 3.3.2.1 Inorganic expansion systems. Inorganic expansion systems based on stabilised acid/base mixtures (essentially carbonate/citrate mixtures) find application in the production of rigid PVC and polystyrene profiles, together with polypropylene sheet. The high permeability of the evolved carbon dioxide through the melt results in increased melt elasticity attributed to cooling of the melt after leaving the die in both free expansion and controlled expansion techniques.

Improvements in the surface coating of the two reactive components appear to have resolved previous problems of pre-decomposition and storage stability.[36]

A secondary, but important, aspect of the improved surface coating is the significant reduction in the sensitivity of temperature control and the screw dwell time, when using these mixtures in comparison with organic expansion systems, where the control of such processing variables remains critical.

Minimum residual odour characteristics have resulted in a growing interest in their use for the production of styrenic and polyolefin films suitable for food contact applications, where tainting may be a problem, (e.g. chocolate packaging). It is however important that the acid/base reaction is complete if problems of surface blooming are to be eliminated.

#### 3.3.2.2 Organic expansion systems. Organic expansion systems are the most widely used, being dominated by sulphonhydrazides and azodicar-bonamide, or its modified grades. It is on the latter type of expansion systems that most developments are currently based. In addition to the

## TABLE 8
### TYPICAL PHYSICAL PROPERTIES OF EXTRUDED FLEXIBLE FOAMS

| Property | Crosslinked LDPE foams | | | | | | Butadiene–acrylonitrile foams[c] | |
| --- | --- | --- | --- | --- | --- | --- | --- | --- |
| | Chemically crosslinked[a] | | | Irradiation-crosslinked[b] | | | | |
| Density (lb ft⁻³) | 2·0 | 4·0 | 6·0 | 2·0 | 3·0 | 6·0 | 2·5 | 16·0 |
| (g cm⁻³) | 0·032 | 0·064 | 0·09 | 0·032 | 0·048 | 0·09 | 0·04 | 0·25 |
| Tensile strength | | | | | | | | |
| (lb in⁻²) | 36 | 81 | 115 | 43 | 65 | 117 | 20 | 200 |
| (kPa⁻¹) | 250 | 360 | 795 | 300 | 450 | 800 | 138 | 1 380 |
| 25% Compression set (%) | 10 | 7 | 5 | 4·9 | 4·5 | 4·2 | 5·0[d] | 20·0[d] |
| Thermal conductivity | | | | | | | | |
| (BTU[e] in⁻¹ ft⁻³ °F⁻¹) | 0·25 | 0·28 | 0·30 | 0·27 | 0·30 | 0·32 | 0·23 | 0·30 |
| (W m⁻² cal⁻¹) | 0·036 | 0·04 | 0·043 | 0·038 | 0·043 | 0·044 | 0·033 | 0·043 |
| Elongation % (Av) | 80 | 100 | 120 | 175 | 215 | 265 | 100 | 250 |

[a] Source: Allen, W. M., *J. Cell Polymers* (Jan./Feb. 1984) 69–72.
[b] Tech. data sheets (Volara), Voltek Inc., Lawrence, Mass., USA.
[c] Ref. 26.
[d] Based on 50% compression set values.
[e] BTU: British Thermal Unit.

developments listed in Table 4, the following grades have recently been introduced:

(1) highly activated grades decomposing in the temperature range 160–165°C;[37]

(2) activated grades with selected particle size range containing additives designed to control crosslinking rates in cellular crosslinked olefins[38] and maximise gas yield.

Developments on other types of expansion systems appear limited although the use of azides as a source of both gas and free radicals in crosslinkable systems has been reported.[39] In addition, an investigation into the decomposition of sulphonhydrazides by exposure to high-frequency sources has also been reported.[40] Both systems indicate novel approaches for the production of cellular extrudates in the future.

### 3.4 Typical Physical Properties

For convenience physical properties, where available, of the types of cellular materials described in this section are summarised in Table 8.

## 4 MISCELLANEOUS FOAM PROCESSING TECHNIQUES

A number of other processing techniques exist for the production of cellular materials. Whilst individually they do not constitute large production units, the resultant products often have properties (e.g. levels of density reduction and the quality of cell structure attained) which cannot be obtained by the use of the more established thermoplastic processing techniques. The two major processing techniques, together with the current level of development, are outlined below.

### 4.1 Press-Moulding Techniques

The production of low-density thermoplastic materials by the press-moulding technique utilising single or two stage expansion techniques could be considered an extension of the rubber processing techniques used for the production of sponge, or closed-cell elastomeric materials. This analogy is further emphasised by the trend towards the production of crosslinked cellular thermoplastics based on olefin/rubber blends. Details of such developments are summarised below, based on individual polymer systems.

### 4.1.1 PVC-Based Systems

As anticipated, two classes of PVC-based foams are commercially available: highly plasticised flexible products produced by a single-stage process, and rigid unplasticised materials produced by a two-stage press-moulding technique.

#### 4.1.1.1 Flexible foams.

These are commercially available in Western Europe, Asia and North America and are mainly produced using the single-stage expansion technique utilising chemical expansion systems, although the use of physical expansion techniques is reported.[41] Variations on the single stage are also apparent; such modification involves the heating of a pre-form at 140–170°C in a pressurised vessel (gas pressure 50–150 psi (0·4–1·0 MN m$^{-2}$)) with expansion occurring on release of the gas pressure.[42]

Considerable latitude in the formulation of a suitable plastisol or plastigel exists. The selection of the PVC polymer grade (i.e. $K$ value 70), plasticiser combination and metal stabiliser will be dependent on the type of chemical expansion system used. The selection is also a function of the processing equipment (e.g. platten temperature and clamping forces available). Expansion systems used are normally dinitrosopentamethylene (DNPT) or highly activated grades of azodicarbonamide (AC), necessitating melt temperatures of 150–180°C to achieve both fusion of the PVC formulation and decomposition of the chemical expansion system.

The two major problems in the production of the press-cured materials are as follows:

(1) Levels of addition for the chemical expansion systems of 10–30 phr result in very high internal mould pressures (4000–5000 psi; 27–34·5 MN m$^{-2}$). High tensile steel moulds, with hardened aluminium, or ground steel gasket/seal plates are essential to eliminate leakage.

(2) the levels of addition of the chemical expansion system pose problems of polymer degradation attributed to the heat evolved during the exothermic decomposition. The selection and level of addition of the thermal stabiliser is therefore important. In addition, if azodicarbonamide is used, the effects of stabiliser choice on the decomposition temperature[43] should be considered.

Typical applications for the flexible closed-cell PVC foams are in buoyancy applications, e.g. life jackets, and specialist anti-vibration applications.

In order to complete this review of flexible PVC foams, particularly thick-sectional products, reference should be made to the elastomer process.[44] This process is used for the production of open-cell, low-density slabstock materials, average thickness (prior to slitting) 5–14 cm, which involves the physical expansion of a plastisol at atmospheric pressure. Carbon dioxide or halohydrocarbons are dissolved in a suitably formulated plastisol under high pressure and at a reduced temperature to maximise gas solubility. The resultant gas-saturated plastisol is pumped onto a continuous silicone rubber bonded glass-fibre belt to give the desired unexpanded thickness. As the wet plastisol progresses away from the applicators towards the fusion units, the temperature of the plastisol increases and the solution pressure decays below the gas solution pressure allowing foaming to begin. The resultant stabilised wet foam is then cured by the use of parallel-plate high-frequency units tuned to match the changing dielectric constant as fusion occurs. Complementary external heat (infrared, or hot air) is necessary to eliminate heat loss from the core of the cellular materials through the cold outer surface of the slabstock foam.

Whilst this process is a unique method for the production of thick-section, low-density, open-cell flexible PVC materials for use in automotive and domestic applications, it is difficult to operate with little flexibility in formulation. Major considerations in the formulation of suitable plastisols are the following.

(1) A low water content in the plastisol is essential to prevent localised heating, or burning in extreme cases. (Moisture levels in the plastisol are controlled between 0·05 and 0·15% water content.

(2) Phthalate-type plasticisers are mainly used to ensure uniformity of heating during both curing and subsequent high-frequency welding operations during fabrication. However, speciality plasticisers are utilised in order to give essential physical properties, e.g. tensile strength, flammability, etc.

(3) Cell stabilisers based on silicones are essential for the stabilisation of the wet foam prior to fusion in the high-frequency units.

*4.1.1.2 Rigid PVC foam.* This foam (density approx. $0·03 \, \text{g cm}^{-3}$) is produced on an international basis and final applications include building panels, refrigerated containers, nautical applications, buoyancy applications and fishing floats suitable for deep water use (not exceeding 100 ft (30 m) depth.

The production techniques involve a two-stage process and are based on one of the following methods:

(1) the use of solvents, acting as fugitive, or temporary, plasticisers;[45]

(2) the use of isocyanates as a means of crosslinking the PVC and as a source of gas during expansion;

(3) the incorporation of a crosslinkable diluent, i.e. a plasticiser, or more commonly a methacrylate monomer[46] utilising azodiisobutyronitrile (AZDN)—also known as azobis isobutyronitrile (AIBN)—as both a polymerisation initiator and a gas source (often in combination with azodicarbonamide).

The formulation and processing of rigid foams are difficult and details of actual production processing techniques are not disclosed, particularly in the production of deep-water floats where surface hardness is of major importance.

Since processing times are long, a balance between the final density and the physical property requirements of the end product is aimed for so as to make the production of such foams economic. As indicated above, the formulations used in these processing techniques contain volatile solvents, within the expansion and crosslinking system, which can result in the emission of potentially toxic residues during processing. These factors, together with the high internal mould pressure, have restricted developments in the area of rigid PVC foams.

Typical physical properties of the rigid PVC foams are given in Table 9.

### 4.1.2 Polyolefin-Based Systems

Whilst non-crosslinked polyolefin foams are in production, limitations in the level of density reduction achievable have resulted in development being concentrated on the area of crosslinked polyolefin foams where higher density reductions can be obtained.

*4.1.2.1 Fundamentals of crosslinked polyolefin foams.* Polyolefins, particularly LDPE, are crystalline polymers, exhibiting very high viscoelastic properties at temperatures below their melting points. At elevated temperatures above the melting point, within the temperature range normally used in the production of cellular materials, a dramatic loss of viscoelastic properties is apparent, preventing the production of extremely low-density materials.

The loss in viscoelasticity is partially overcome by the use of chemical or irradiation crosslinking techniques, resulting in an increased melt elasticity, sufficient to accept the extremely high rate of extension of the melt without rupture of the cell walls occurring during the expansion stage.

Comparative details of the melt elasticity of crosslinked and non-crosslinked polymers melts are shown schematically in Fig. 2.

TABLE 9

TYPICAL PHYSICAL PROPERTIES OF RIGID CLOSED-CELL PVC FOAMS

| Density (average) | | | | | | | |
|---|---|---|---|---|---|---|---|
| Density (average) | $(kg\,m^{-3})$ | 40 | 55 | 75 | 100 | 225 | 340 |
| | $(lb\,ft^{-3})$ | 2·5 | 3·44 | 4·7 | 6·25 | 14·0 | 21·25 |
| Comprehensive strength | $(kPa)$ | 350 | 550 | 950 | 1400 | 5000 | 8200 |
| | $(lbf\,in^{-2})$ | 50 | 77 | 135 | 200 | 725 | 1190 |
| Compressive modulus | $(kPa)$ | 10500 | 17500 | 30000 | 40000 | 80000 | 110000 |
| | $(lbf\,in^{-2})$ | 1500 | 2500 | 4300 | 5700 | 10900 | 15900 |
| Shear strength | $(kPa)$ | 350 | 490 | 950 | 1160 | 1650 | 1725 |
| | $(lbf\,in^{-2})$ | 50 | 70 | 135 | 165 | 240 | 250 |
| Flexural strength | $(kPa)$ | 450 | 700 | 1050 | 1400 | 3000 | 4000 |
| | $(lbf\,in^{-2})$ | 65 | 100 | 150 | 200 | 435 | 580 |
| Flexural modulus | $(kPa)$ | 15000 | 20000 | 45000 | 50000 | 100000 | 120000 |
| | $(lbf\,in^{-2})$ | 2175 | 2900 | 6525 | 7250 | 14500 | 17400 |
| Tensile strength | $(kPa)$ | 500 | 800 | 1200 | 1600 | 2500 | 3370 |
| | $(lbf\,in^{-2})$ | 72 | 116 | 174 | 230 | 360 | 490 |
| Buoyancy | $(kg\,m^{-3})$ | 945 | 915 | 895 | 880 | 730 | 685 |
| | $(lb\,ft^{-3})$ | 59 | 57 | 56 | 55 | 46 | 39 |

|  | | | | | | |
|---|---|---|---|---|---|---|
| Thermal conductivity at 10°C (50°F), BS 4370 Part 2 ($W\,m^{-1}\,°C^{-1}$) | 0·027 | 0·033 | 0·038 | 0·040 | 0·042 | 0·044 |
| Coefficient of linear thermal expansion ($°C^{-1}$) | $35 \times 10^{-6}$ | $35 \times 10^{-6}$ | $35 \times 10^{-6}$ | $.35 \times 10^{-6}$ | $35 \times 10^{-6}$ | $35 \times 10^{-6}$ |
| Water vapour permeability at 23°C and 85% RH, ISO 1663 ($mg\,m^{-2}\,s^{-1}$) | 150 | 100 | 80 | 80 | 80 | 80 |
| Water absorption (% w/v) | 0·5 | 0·5 | 0·5 | 0·5 | 0·5 | 0·5 |

Fire resistance — BS 476 Part 7 Class 1
BS 3869 self-extinguishing, will not support combustion

Max. operating temperature (bare metal) — Continuous use 60°C (140°F)
Intermittent use 80°C (176°F)

Chemical resistance — Resistant to aliphatic hydrocarbons, dilute acids and alkalis, sea water, methylated spirits, fuels containing small quantities of aromatic hydrocarbons (such as petrol and diesel), polyester resin with low styrene content
Attacked by: ketones, aliphatic esters, chlorinated hydrocarbons
Slightly softened by aromatic hydrocarbons

Data source: Tech. Data Sheet TS1081 Plasticell, Permali Gloucester Ltd, UK.

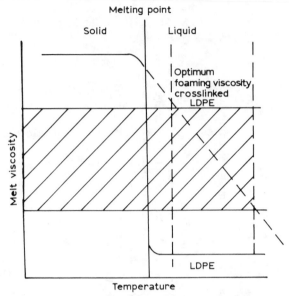

FIG. 2.    Comparative melt viscosity of polyethylene.

*4.1.2.2 Selection of polyolefin type.* Optimisation of the melt elasticity properties within the temperature range used during foaming can be achieved by the selection of a suitable polymer system. Considerable information has been published on the selection of suitable polyolefins for use in foam processes.[47] The molecular structure of the olefin homo- or co-polymer is of importance particularly in chemical crosslinking processes. The structural requirements of suitable polymers can be briefly summarised as follows.

(1)  polymers with a high level of crystallinity (i.e. increasing density) will not give the necessary high crosslink density;

(2)  polymers with high levels of branching give high crosslink densities, but long side chains are reported to inhibit crosslinking during peroxide curing;[48]

(3)  copolymerisation or the physical blending of polymers to increase the degree of unsaturation, e.g. ethylene vinyl acetate copolymer, or blends of LDPE/butyl rubber, LDPE/EPDM, both increase crosslink density and improve physical properties in the end product;

(4)  the addition of reactive co-monomers, e.g. acrylate monomers, is also being investigated.

## 4.2 Developments in Rotational Moulding

Development in the areas of rotational moulding and 'expansion casting' processes, which offer considerable advantages over injection moulding in the production of large hollow parts, has been at a low level. Interest in the concept of cellular rotational moulding increased during the 1980s and was the subject of limited development by both the polymer producers and the suppliers of chemical expansion systems,[49] particularly in the USA. However, only limited commercial production of cellular rotational moulded parts is currently known to exist in Western Europe.

Main considerations in the development of cellular rotational moulding include selection of materials, processing techniques, equipment and processing variables.

### 4.2.1 Material Developments

The selection of both expansion system and polymer grade are more critical than in conventional cellular thermoplastic processing and necessitate a particular mention.

#### 4.2.1.1 Expansion systems.
The dispersion of the chemical expansion system in the polymer poses problems, as poor dispersion will give problems of blistering, or irregular cell structures. The preparation of melt compounds using conventional extrusion compounding techniques and subsequent cryogenic grinding is therefore preferred to the use of simple powder blending techniques. However, a careful control over both powder size and shape is still necessary to retain good powder flow properties during processing.

The selection of the chemical expansion system used is normally restricted to the lower-temperature expansion systems, e.g. oxy-bis(benzenesulphonhydrazide), or highly activated grades of azodicarbonamide, and is a function of the polymer type to be expanded and processing conditions.

#### 4.2.1.2 Type of polymer used.
Polyolefins have been used predominantly in cellular rotational casting, although in addition polystyrene and ABS have been used in the production of rigid mouldings using the expansion casting techniques.[50] Minimal development work has been carried out on plasticised PVC due to major problems with mould extraction from both copper and phosphated alloy tools. In addition, the low tear strength of hot cellular plasticised PVC also gives mould extraction problems. These problems do not make cellular flexible PVC rotational moulding a commercially viable proposition.

*4.2.1.3 The crosslinking system.* The selection of a suitable crosslinking system can prove problematic due to the compounding heat history involved. Organic peroxides with a short half-life are necessary to ensure rapid crosslinking of the olefin matrix, but must survive melt temperatures of 100–115°C during the compounding of both the peroxide and chemical expansion systems if the necessary levels of dispersion are to be achieved. As a result most formulations are based on dicumyl peroxide, or 2,5-dimethyl-bis(5-butylperoxy)hexane, both of which satisfy the performance requirements of the process.

Obviously the above comments relate only to chemical crosslinking techniques. When irradiation crosslinking techniques are employed higher compounding temperatures can be used with decomposition of the expansion system being the controlling factor. Crosslink density is a direct function of dwell time/radiation dosage, relative to the dimensions of the unexpanded blank. Increased radiation dosage will result in degradation of the polymer surface due to reversion, and consequent loss of gas from the melt during expansion.

### 4.2.2 Processing Techniques

Two methods of production are known for the production of low-density chemically crosslinked polyolefin foams.

*4.2.2.1 Single-stage curing systems.* Single-stage press curing operations are the most widely used processing techniques, but, whilst satisfactory for the production of medium-density crosslinked ethylene vinyl acetate sheet for shoe soling applications, problems occur in the production of low-density foams.

The problems arising are essentially mechanical and arise in the selection of the press and in mould design. Major considerations include the following.

(1) Multi-platten (multi-daylight) press units are suitable for the production of medium-density ethylene vinyl acetate/LDPE foams ($0.8\,\mathrm{g\,cm}^{-3}$) where internal mould pressures are as low as 1000–1500 psi ($7–10\,\mathrm{MN\,m}^{-2}$) and the volume expansion on extraction from the mould is not large. Such processes are economic to operate, permitting the maximum utilisation of the platten surface without the necessity for high clamping forces.

(2) For the production of low-density foams ($0.07\,\mathrm{g\,cm}^{-3}$) it is necessary to utilise press units modified to give extremely high opening speeds, e.g. bottom-operating presses, or press units

permitting the use of reverse hydraulic platten opening cylinders. The rate of expansion and final volume of the expanded moulding after ejecting from the mould cavity require virtually instantaneous opening of the mould platten to prevent the edge splitting. Practically this process is not economic to operate since the total platten area cannot be used and any contact of foam with the mould surface or a supporting pillar will result in tearing of the mouldings.

(3) As indicated previously, high internal mould pressures (4000–5000 psi; 27–34·5 MN m$^{-2}$) necessitate the use of high-tensile steel moulds with suitable gaskets. Moulds with 'draw angles' of 30°, with minimal undercut areas, are essential to minimise edge splitting and tearing during ejection from the mould cavity. In practice scrap levels below 9–10% have not been achieved; together with the problems of press design, this has restricted development in this area of processing.

*4.2.2.2 Two-stage curing system.* The two-stage press curing techniques involving partial crosslinking of the polyolefin and decomposition of the expansion system in the mould, followed by post-expansion in a hot air oven or a liquid cooling bath, are the subject of current development. Process control is critical since the balance between crosslinking and expansion during the press cure cycle will control both final density and surface quality during the post-expansion stage. Advantages of this process, analogous to the two-stage rigid PVC process, are

(1) the total utilisation of the platten area is possible;
(2) foams with cell quality and density reduction similar to those produced using the conventional pressure moulding process are obtainable.

A further development of the two-stage process is a patented process involving the use of irradiation crosslinking and gas saturation prior to post-expansion. A control over cell diameter and physical properties is claimed for this process, but details of the processing technique are unavailable, particularly the method for controlling cell size.

Details of both formulation and typical physical properties of crosslinked olefin foams are given in Table 10.

*4.2.3 Equipment Developments*
Two major developments have occurred in the development of mould design since the initial concept of cellular rotational moulding in the late 1960s; these are as follows.

## TABLE 10
FORMULATIONS AND PHYSICAL PROPERTIES OF CHEMICALLY CROSSLINKED, PRESS CURED, POLYOLEFIN FOAMS

| Formulation | 1 | 2 | 3 | 4 | 5 | 6 |
|---|---|---|---|---|---|---|
| 'Evatane' EVA copolymer 538 | 100 | 100 | 100 | 80 | 80 | 75 |
| 'Evatane' EVA copolymer 28-05 | — | — | — | — | — | — |
| 'Evatane' EVA copolymer 33-25 | — | — | — | — | — | — |
| 'Alkathene' polyethylene 11 | — | — | — | 20 | 20 | 25 |
| 'Retilox' F40[a] | 5 | — | 2 | 2 | 2 | 5 |
| 'Perkadox' 14/40[b] | — | — | — | — | — | — |
| Dicumul peroxide | — | 5 | — | — | — | — |
| Azodicarbonamide | 5 | 5 | 5 | 5 | 5 | 5 |
| Calcium carbonate (whiting) | — | — | — | — | 20 | 20 |
| 'Winnofil' S[c] | — | — | — | — | — | — |
| 'Ultrasil' VN3[c] | — | — | — | — | — | — |
| 'Silteg' AS7[c] | — | — | — | — | — | — |
| Zinc oxide | 2·5 | 2·5 | 2·5 | 2·5 | 2·5 | 2·5 |
| Stearic acid | 0·5 | 0·5 | 0·5 | 0·5 | 0·5 | 0·5 |
| Properties | | | | | | |
| Cure temperature (°C) | 165 165 | 165 | 165 165 | 165 165 | 165 165 | 165 |
| Cure time (min) | 8 10 | 8 | 8 10 | 10 15 | 10 | 10 |
| Density (g cm⁻³) | 0·12 0·14 | 0·11 | 0·08 0·08 | 0·08 0·07 | 0·09 | 0·18 |
| Shore A hardness | 38 40 | 34 | 20 25 | 28 33 | 27 | 50 |
| Split tear strength (kg) | 4·1 2·7 | 2·9 | 4·1 3·9 | 3·4 2·6 | 4·6 | 4·2 |
| Elongation at break (%) | 205 105 | 125 | 260 250 | 230 180 | 260 | 170 |
| Compression set[d] (%) | | | | | | |

Data source: ICI Technical Data Sheet.[50]
[a] 'Retilox' F40 is supplied by Montedison (UK) Ltd.
[b] 'Perkadox' 14/40 is supplied by Akzo Chemie (UK) Ltd.
[c] 'Winnofill' S, 'Ultrasil' and 'Silteg' AS 7 are supplied by ICI Mond Division.
[d] In this test the sample has been compressed under a load of 7 kg cm⁻² for 24 h at room temperature. The set was measured 1 h after removal of the load. The information given for formulations 11–19 has been provided by ICI Mond Division.

*4.2.3.1 Single-layer rotationally moulded components.* Single-layer rotationally moulded parts are the most developed, but pose problems of poor surface finish, partially offset by reduced polymer costs and increased rigidity in the moulded parts. Mould designs are similar to those used for conventional rotational moulding, but draw angles are increased to compensate for extraction problems with rigid-walled mouldings. Attention is also given to increasing the uniformity of heat transfer over the mould surface to reduce problems of irregular cell structures, or localised predecomposition.

TABLE 10—contd.

| 7 | 8 | 9 | 10 | 11 | 12 | 13 | 14 | 15 | 16 | 17 | 18 | 19 |
|---|---|---|----|----|----|----|----|----|----|----|----|----|
| 75 | 75 | — | — | 100 | 100 | 100 | 100 | 100 | 100 | 100 | 100 | 100 |
| — | — | 100 | — | — | — | — | — | — | — | — | — | — |
| — | — | — | 100 | — | — | — | — | — | — | — | — | — |
| 25 | 25 | — | — | — | — | — | — | — | — | — | — | — |
| 2 | 5 | 2 | 2 | — | — | — | — | — | — | — | — | — |
| — | — | — | — | 3 | 3 | 3 | 3 | 3 | 3 | 3 | 3 | 3 |
| 5 | 4 | 5 | 5 | 4 | 4 | 4 | 4 | 4 | 4 | 4 | 4 | 4 |
| 20 | 20 | — | — | — | — | — | — | — | — | — | — | — |
| — | — | — | — | — | 12·5 | 25 | 50 | — | — | — | — | — |
| — | — | — | — | — | — | — | — | 12·5 | 25 | — | — | — |
| — | — | — | — | — | — | — | — | — | — | 12·5 | 25 | 50 |
| 2·5 | 2 | 2·5 | 2·5 | 2 | 2 | 2 | 2 | 2 | 2 | 2 | 2 | 2 |
| 0·5 | 0·5 | 0·5 | 0·5 | 0·5 | 0·5 | 0·5 | 0·5 | 0·5 | 0·5 | 0·5 | 0·5 | 0·5 |
| 165 | 165 | 165 | 165 | 160 | 160 | 160 | 160 | 160 | 160 | 160 | 160 | 160 |
| 10 | 10 | 10 | 10 | 15 | 15 | 15 | 15 | 15 | 15 | 15 | 15 | 15 |
| 0·09 | 0·25 | 0·08 | 0·08 | 0·17 | 0·21 | 0·22 | 0·23 | 0·34 | 0·5 | 0·32 | 0·37 | 0·38 |
| 27 | 64 | 20 | 20 | 32 | 35 | 40 | 39 | 65 | 80 | 48 | 69 | 69 |
| 4·4 | 7·8 | — | — | 2·5 | 3·7 | 4·3 | 4·7 | 6·5 | 10·1 | 5·7 | 8·5 | 7·0 |
| 250 | 280 | — | — | 160 | 190 | 190 | 190 | 300 | 280 | 220 | 240 | 130 |
| — | — | — | — | 25 | — | — | — | — | 5 | 4·5 | 6 | 3 | 4 |

*4.2.3.2 Multi-layer rotationally moulded components.* Multi-layer rotationally moulded parts, whilst offering advantages of improved surface quality and, by the selection of polymer type and individual layer thickness in the resultant laminate structure, giving enhanced physical properties, pose problems of mould design. With the necessity of introducing up to three consecutive polymer charges into the mould cavity, sophisticated metering/feed mechanisms are required. A range of feed mechanisms have been suggested[51,52] involving radio-controlled containers within the mould cavity, or lance injection systems utilising a multi-station oven. However, the multi-layer moulding technique requires considerably more development in both process and mould design to achieve a commercially viable process.

### 4.2.4 Processing Variables

Limited technical information has been published on the production of single-, two- and three-layer mouldings, mainly due to the necessity to establish processing conditions for individual mouldings. Major considerations include the following.

(1) On the basis of wall thickness and configuration of the cellular moulding, comparative cycle times for cellular parts are longer than those for similar solid mouldings using similar polymers and loadings.

(2) The level of blowing agent addition is directly related to cycle time. Increased levels of blowing agent addition will reduce oven dwell times, as a result of the heat evolved during the exothermic decomposition, contributing to the total heat input necessary for fusion of the polymer system.

Typical moulding cycles illustrating the use of azodicarbonamide and oxy-bis(benzene sulphonhydrazide) are shown in Table 11, which indicates the complexity of the process.

### 4.2.5 Typical Physical Properties

Due to the limited development in this area of processing, few published physical test data are available. Typical properties for a LDPE foamed rotationally moulded part are shown in Table 12, but refer only to the single-layer process. Considerable improvements in impact properties and deflection loading will be apparent in both two-layer and three-layer structures, assuming that good adhesion between consecutive layers is obtained.

Surprisingly, considerably more information on the physical properties of chemically crosslinked rotationally moulded foams is available.[53,54]

### 4.2.6 Discussion

Little further process development is anticipated in the area of cellular rotational moulding, although the process constitutes the most economic method for the production of very large cellular items, particularly hollow mouldings, e.g. containers, etc.

Limitations in part design and long cycle times, whilst offset by low mould construction costs, do not make the process economic by comparison with other cellular thermoplastic processing techniques.

## TABLE 11
### DATA FROM ROTATIONAL MOULDING TRIALS

| Polymer | CBA | Oven Time (min) | Oven Temp. (°F) | Air time (min) | Water cool (min) | Comments |
|---|---|---|---|---|---|---|
| MDPE, MFI = 3·0 | Celogen® AZ | 9 | 500 | 2 | 5 | SG = 0·335. Very thick part |
| HDPE, MFI = 7·0 | Celogen® OT | 12·5 | 600 | 8 | 2 | Judged to be a good part; AZ did not work here |
| MDPE, MFI = 3·0 | Celogen® AZ | 20 | 500 | 4 | 4 | SG = 0·471 |
| MDPE, MFI = 3·0 | Celogen® AZ | 15 | 650 | 2 | 10 | Well foamed part. SG = 0·451 |
| MDPE, MFI = 3·0 | Celogen® OT | 12 | 600 | 2 | 10 | Well foamed and thick part; somewhat under-cured |
| HDPE, MFI = 4·0 | Celogen® OT | 14 | 525 | 5 | 12 | Good part. SG = 0·541 |

Source: *Journal of Cellular Plastics* (USA) March/April 1983.[52]
® Trade mark of Uniroyal Chemical Co., USA.

## TABLE 12
TYPICAL PHYSICAL PROPERTIES OF FOAM ROTATIONALLY MOULDED 8305[a]

| Property | | Celogen® AZ130[c] (%) | | | | |
|---|---|---|---|---|---|---|
| | | 0 | 0·2 | 0·5 | 0·8 | 1·0 |
| Density (g cm$^{-3}$) | | 0·88 | 0·55 | 0·38 | 0·33 | 0·28 |
| Thickness (in) | | 0·144 | 0·242 | 0·345 | 0·406 | 0·481 |
| Impact at $-20$°C | Mould side | 14·5[b] | 12·4 | 4·8 | 3·3 | 2·6 |
| (ft lb in$^{-2}$) | Inside | 16·0 | 7·3 | 3·9 | 2·8 | 3·1 |
| Crush (lb) | | 1 500 | 1 820 | 2 250 | 2 520 | 2 450 |
| Stress crack (h) | Mould side | | No failures after 300 h | | | |
| | Inside | 360 | 340 | 50 | 50 | 50 |

Note: The above data should only be compared with the control sample, i.e. 0%
  Celogen AZ, and not with other physical property data related to 8305.
[a] SCLAIR 8305, LDPE resins ex DuPont Canada Inc.
[b] The five samples tested flexed without breaking.
[c] Celogen is the trade mark for a Uniroyal Chemical blowing agent.
Raw materials/processing conditions
Polymer system: SCLAIR 8305, density, 0·932 g cm$^{-3}$; Melt Flow Index 3·0.
Expansion system: Celogen AZ130; azodicarbonamide.
Mould dimensions, aluminium base mould: $11\frac{1}{2} \times 11\frac{1}{2} \times 5\frac{3}{4}$ in.
Charge weight: 40 oz.
Cycle times:
  (a)  17 min at 650°F (330°C) for azodicarbonamide loadings of 0·2% and 0·5%;
  (b)  15 min at 650°F (330°C) for azodicarbonamide loadings of 0·8% and 1·0%;
both followed by 2 min air cooling, followed by 10 min water cooling at 80°F (25°C).


## 5 CONCLUSIONS AND FUTURE DEVELOPMENTS

The production of cellular thermoplastics is an established sector within
the total polymer processing industry.

Whilst most development activity is currently centred on injection
moulding, recent developments have been extended to include the areas of
extrusion and press moulding techniques.

The impetus for this development, involving the polymer producers, the
additive suppliers and the machinery manufacturers, originate from two
needs:

  (1)  to improve surface quality, reduce flammability and increase the
       reproducibility of physical properties in mouldings to make such
       products competitive with homogeneous mouldings;

(2) to develop extrusion and press cure processing techniques capable of producing low-density materials with fine uniform cell structures giving physical properties not attainable by other foaming techniques.

Obviously these performance characteristics are highly desirable independently of the processing technique being used. It is therefore suggested that current developments will overlap, resulting in mouldings and/or extrudates with reproducible physical properties, e.g. low densities and uniform cell structures, and outer surfaces requiring minimum surface finishing operations, to give products which can successfully compete with homogeneous materials in the new application areas.

## REFERENCES

1. BAYER, C. E. *et al.*, US Patent No. 3058161 to Dow Chemicals, USA, Oct. 16th 1962.
2. PLASKAN, L. F., Structural foam—approach it as a system. *Plastic World*, **35** (Sept. 1977) 38–41.
3. THRONE, A. J., Structural foam moulding parameters. *J. Cell Plastics*, May/June (1976) 161–76.
4. HENDRY, J., Structural foam processing affiliations, yesterday, today and tomorrow. *J. Cell Plastics*, July/Aug. (1979) 220–2.
5. THRONE, A. J., The new structural foam techniques. *J. Cell Plastics*, Sept./Oct. (1976) 263–84.
6. HAMEL, G. & GRIBENS, J. A., Moulding cycle optimisation of the structural foam process. *Transactions of the SPE ANTEC Conference*, Society of Plastics Engineers, Connecticut, USA, 1985, pp. 447–9.
7. GROSS, L. H. & ANGELLI, R. G., Boon to low pressure structural foam moulding—'Swirl free foam parts'. *Plastics Technology*, May (1976) 33–6.
8. TRANQUILLA, M. N., HAZNECI, N. & HENDRY, J., Designing, prototyping, tooling and processing of large structural foam parts. *J. Cell Plastics*, May/June (1981) 152–68.
9. STERNFIELD, A., Where are they using sandwich moulding and why. *Modern Plastics International*, June (1983) 25–7.
10. SANDIFORTH, D. J. H. & OXLEY, D. F., Integral skin foam moulding. *Plastics & Polymers*, **39** (1972) 288.
11. ECKARDT, H., Co-injection, charting new territory and opening new markets. *J. Cell Plastics*, **123** (Dec. 1987) 555–92.
12. MEYER, W., Do advanced structural foam moulding processes reduce cost? *J. Cell Plastics*, Jan./Feb. (1978) 73–5.
13. SNELLER, J., Thin walling with structural foam, low cost route to better parts. *Mod. Plastics International*, Aug. (1980) 24–6.
14. Technical data sheet, Hettinga Equipment Inc., University Av Des Moins, Iowa 50311, USA, Oct. 1986.

52 K. T. COLLINGTON

15. Injection moulded microcellular foams. Provisional data sheet for alkathene EVA copolymer compounds 0496, ICI Polyolefins Div., Wilton, Middlesbrough, UK.
16. SHINA, N., TSUCHIYA, M. & NAKAE, H., Properties and application techniques of crosslinked polyethylene foams (2). *Japan Plastics Age*, Jan. (1973) 49–53.
17. Structural foam is the key to modernised bed design (editorial report). *Transactions of Plastics Design Forum*, July/Aug. (1978) 82–5.
18. KIRKLAND, C., New injection and foam materials debut at the Structural Foam Confab. *Plastics Technology*, June (1985) 13–19.
19. Production of foam plastics with Hydrocerol-compound. Technical data sheet, Boehringer Ingelheim, Chemicals Div., W. Germany.
20. Lower density, higher speeds wider structural foam horizons (editorial). *Modern Plastics International*, Dec. (1978) 62.
21. HAZNECI, N., Manufacturing economics favour structural foams in business machines. *Plastics Technology*, Dec. (1978) 75–80.
22. BARTH, H. J., Extrusion dies for rigid PVC foam profiles. *Eur. J. Cell Plastics*, July (1979) 103–9.
23. KIESLING, G. C., Extruded structural foam. *J. Cell Plastics*, Nov./Dec. (1976) 337–40.
24. BUSH, F. R. & ROLLFSON, G. C., The ABC of co-extruding foam-core ABS pipe. *Modern Plastics International*, Feb. (1981) 38–40.
25. KIRKLAND, C., Crosslinked PE foam sheet—New continuous process arrives. *Plast. Technology*, Nov. (1980) 89–92.
26. McCRACKEN, W. J., Nitrile/PVC and other polymer/resin closed cell foams. *J. Cell Plastics*, March/April (1984) 150–6.
27. PURI, R. R. & COLLINGTON, K. T., The production of cellular crosslinked polyolefins—Part 1, *Cellular Polymers*, May (1988) 56–84.
28. RIDDLE, M. J., The crosslinkability of LDPE resins. *Transactions of the SPE ANTEC Conference*, New York, Society of Plastics Engineers, Connecticut, USA, 1980, pp. 183–7.
29. RADO, R., SIMONIK, J. & SUCHA, H., Expansion and network crosslinking of branched polyethylene in a continuous process Part 1 and Part 2. *Plasty and Kucuk*, **16** (1979) No. 2, 38–42 and No. 4, 103–6.
30. NORMANTON, J. K., Extrusion of telephone cable insulation using expandable medium density polyethylene compounds. *Transactions of the Int. Wire and Cable Symposium*, 4–6 Dec. 1974, Cherry Hill, NJ, USA, pp. 182–7.
31. COLLINGTON, K. T., Extrusion of cellular thermoplastics. In *Developments in Plastics Technology—1*, ed A. Whelan & D. Dunning, Applied Science Publishers, London, 1982, p. 45–74.
32. SZAMBORSKI, E. C. & MARCELLI, R. A., Designing with plastics—rigid foamed—PVC pipe, *Plastics Engineering*, Nov. (1976) 49–51.
33. See Ref. 26.
34. HODGSON, T. C., An instrument for the analytical control of the chemical blowing of cellular polymers. *J. Cell Plastics*, Nov./Dec. (1973) 148–57.
35. ORIMO, K. & YAMAMOTO, S., Gas injection extrusion process for a foamed plastic insulation supplying a fixed quantity of gas. *Transactions of the Int. Wire & Cable Symposium*, 4–6 Dec. 1974, Cherry Hill, NJ. USA.
36. See Ref. 18.

37. Technical data sheets. Schering Industrial Chemicals, Cambridge. Ref. number HGE/6b and HGE/7b.
38. SIMONIKOVA, J., SVBODA, J. & SIMONIK, J., The use of chromium modified azodicarbonamide as a blowing agent for compound foaming. *J. Cell Polymers*, 5(3) (1986) 309–21.
39. MUENCHOW, J., POSTURINO, R., HALLE, R. & LEWIS, R., A crosslinking/foaming agent for polyethylene, *Transactions of the SPE ANTEC Conference*, Society of Plastics Engineers, Connecticut, USA, 1986, pp. 692–4.
40. MENGES, G. & BEISS, K., A new process for polyethylene foaming. *Transactions of the SPEC ANTEC Conference*, Society of Plastics Engineers, Connecticut, USA, 1986, pp. 686–8.
41. COLLINGTON, K. T., Cellular PVC, Developments in PVC Productions and Processing, Applied Science Publishers, London, 1977, pp. 109–25.
42. See Ref. 41.
43. ROWLAND, D. G., Activators control gas evolution of blowing agents. *Rubber & Plastics News*, (21 Sept. 1987) 14–17.
44. US Patent 2666036 to Elastomer Chemical Corp., Newark, USA.
45. BP Patents 714606 and 716366, 1954, Thomas Whiffen & Son, UK.
46. MAO, C. & VERNON, M., Application of the synergy of two blowing agents in the fabrication of rigid PVC foams. *J. American Chem. Soc.*, 135(1) (1986) 51–4.
47. See Ref. 28.
48. See Ref. 29.
49. New versatility for rotomoulding with new materials, new equipment design concepts (editorial). *Mod. Plastics International*, Dec. (1978) 12–14.
50. Sinter casting. ICI Technical data sheet, 1972.
51. HECK, R. L., Improved technology of rotationally moulded foam finds new applications. *Plastics Engineering*, Nov. (1982) 37–9.
52. KRAVITZ, H. & HECK, R. L., Now's the time to look into foam. *Rotational Moulding, Plastics Technology*, Oct. (1979) 63–6.
53. NORKIS, M., MILTZ, J. & PAUKEN, L., Properties of rotational—moulded crosslinked polyethylene foams. *J. Cell Plastics*, Nov/Dec. (1975) 323–7.
54. REES, R. L., Polyethylene and crosslinked polyethylene for rotational moulding, *Transactions of the SPE ANTEC Conference*, Society of Plastics Engineers, Connecticut, USA, 1981, pp. 621–3.

Chapter 2

# RECENT ADVANCES IN ANALYSIS AND CHARACTERIZATION OF POLYMERS AND PLASTICS

MASAO TAKAHASHI

*Toray Techno Co. Ltd, Otsu, Japan*

## 1 INTRODUCTION

Very remarkable developments in polymer science and technology are continuously being made where the introduction of these new polymeric materials has taken place with considerable interest. Examples include high-temperature or fully aromatic polymers such as poly(phenylene sulfide), aromatic polyesters and polyamides, polyimides and poly-(arylether ether ketone), conducting polymers such as polyacetylene, polypyrrole, polythienylene and polymeric bridged macrocyclic transition-metal complexes, ultra-high-molecular-weight polyethylene, various fluoropolymers and polymers based on new concepts such as interpenetrating network polymers and liquid-crystalline polymers.

Applications of polymers have been widely diversified in many industrial and practical fields and a number of speciality polymers have been developed to meet special needs. A variety of polymer blends and alloys have been developed in practical applications in the plastics and elastomer fields. Advanced composites require high-performance matrix polymers and polymer compositions. The analysis and characterization of polymers and plastics are important in all the aspects described above. A number of studies have been reported in relation to them.

Tools and techniques with high performance for analysis and characterization have been developed. A number of publications and presentations related to these developments as well as their applications to polymer analysis and characterization can be found in many journals and conference proceedings. More than 400 literature citations can be made for the last three years.

These tools and techniques for polymer analysis and characterization are summarized as follows.

(1) Molecular characterization by means of solution characterization.

    (a)   Measurement of average molecular weight.

    (b)   Measurement of molecular weight distribution.

    (c)   Measurement of the dimensions and rigidity of the polymer molecular chain.

    (d)   Long chain branching.

    (e)   Copolymer composition.

(2) Characterization by thermal techniques.

    (a)   Differential scanning calorimetry.

    (b)   Differential thermal analysis.

    (c)   Thermogravimetry.

    (d)   Thermomicrophotometry.

    (e)   Thermomechanical analysis.

(3) Electron microscopy method for characterization.

    (a)   Scanning electron microscopy (SEM). The depth of focus is very large and higher resolution can be obtained without great skill. Sample preparation is relatively easy and observation of samples having very complex shapes and contours can be readily carried out. By combining the X-ray microanalyzer (XMA), elemental analysis in a small area becomes possible.

    (b)   Transmission electron microscopy (TEM). The resolution of TEM has been remarkably improved and has recently reached levels of 1–3 Å. By using high-resolution TEM (HRTEM), structures of molecular and atomic levels can be determined and very fine structural analysis is thereby possible. The XMA technique can be used with TEM and its spatial resolution is 100–200 Å, which is much higher than that of SEM-XMA. Elemental and state analysis can be done by combining electron energy loss spectroscopy (EELS) with scanning TEM (STEM). The sample preparation technique is vital for TEM and techniques for preparing ultra-thin samples as well as giving higher contrast to the samples are reported.

(4) Analytical pyrolysis of polymers.

(5) Spectroscopic methods for analysis and characterization.
    (a) Infrared spectroscopy (especially FT-IR). Fourier transform infrared spectroscopy (FT-IR) having high sensitivity and accuracy and large data processability is very highly effective for polymeric materials which usually have very complicated structural and compositional features. New techniques that have been developed include FT-IR-DRIFT (diffuse reflective infrared-FT), -ATR (attenuated total reflectance), -RAS (reflection, absorption spectroscopy), -EMS (emission spectroscopy), -PAS (photoacoustic spectroscopy), -SEWS (surface electromagnetic wave spectroscopy) and -microscopy, which make FT-IR much more important than before. Techniques such as partial enlargement of the spectrum, difference spectra and curve fitting are also very useful. FT-IR is also being widely used for surface analysis and micro-area analysis.
    (b) Raman spectroscopy. Since lasers have been used for the light source of Raman spectroscopy, this technique has been becoming one of the widely used analytical and characterization tools for polymeric materials. Measurements on the far-infrared region can be continuously carried out without the water and glass container interfering with them. Microprobe, ATR and SERS (surface-enriched Raman scattering) techniques have been recently developed.
    (c) X-ray photoelectron spectroscopy (XPS or ESCA). XPS is a non-destructive method and gives clear information about the chemical composition, the state and the structure on the surface of the material examined. The depth of the information reaches only about 5–20 Å and therefore it is very effective for surface analysis.
    (d) Nuclear magnetic resonance (NMR)—especially FT-NMR and high-resolution NMR. The sensitivity of NMR has made great strides by utilizing the ultra-conductive magnet and the pulse-FT method. $^{13}C$-NMR, which can directly measure carbon atoms, the basic structural element of polymers, and measurement of solid polymers by means of HR-NMR (high-resolution NMR) have thereby become possible
        Chemical information on the polymer such as the basic chemical structure, type of end groups, the composition and the monomer sequence distribution of copolymer and spatial regularity can all be obtained by means of NMR.

(e) Secondary-ion mass spectrometry (SIMS). SIMS has recently become a more useful tool for the analysis and characterization of polymers, by utilizing its several potential advantages such as molecular specificity, surface sensitivity and high spatial resolution. However, several practical problems still exist and many reports have appeared mainly through SIMS international conferences.

(6) X-ray diffraction and scattering.

(a) X-ray diffraction. This method has been widely used for many years as one of the basic techniques for polymer characterization. Fine structures and crystal structure analysis of polymers have been thereby carried out.

(b) X-ray small-angle scattering. This has been recently developed for characterization of polymer fine structure where the strong X-ray sources, highly sensitive detectors and high-speed multi-channel analyzers have been developed so that the time required for the measurement has been considerably shortened to the order of minutes. This technique is especially useful for analyzing the phase and interface structures of polymer systems.

In this chapter, recent advances in the molecular characterization by means of solution and thermal techniques, based on the literature, will be reviewed.

In many industrial applications such as composites, coatings, adhesives, electronics, etc., problems related to surfaces and interfaces are very often important. Science and technology for the analysis and characterization of surfaces have shown remarkable developments.

Four main effective spectroscopic methods for analysis and characterization of surfaces are electron-probe micro-analysis (EPMA), Auger electron spectroscopy (AES), secondary-ion mass spectrometry (SIMS) and X-ray photoelectron spectrometry (XPS or ESCA).

The surface area and depth to be analyzed by each method are as follows.

|      | Area diameter (mm) | Depth ($\overset{\circ}{A}$) |
|------|--------------------|------------------------------|
| EPMA | 0·001–0·3          | $10^4$                       |
| AES  | 0·1–1              | 10                           |
| SIMS | 0·001–1            | $10^7$                       |
| XPS  | 1–3                | 5–20                         |

On the other hand, a number of unique developments related to the analysis and characterization of surface have also been achieved by means of infrared and Raman spectroscopy, which are usually effective for analysis and characterization of bulk polymer. In this chapter, as space is limited, some of the recent developments in polymer surface analysis and characterization by means of only spectroscopic methods such as XPS, SIMS, IR and Raman will be reviewed.

## 2 SOLUTION CHARACTERIZATION BY GEL PERMEATION CHROMATOGRAPHY/LOW-ANGLE LASER LIGHT SCATTERING (GPC/LALLS)

As Hjertberg et al. have described,[1] gel permeation chromatography (GPC) was a breakthrough in the characterization of polymeric materials where the relative molecular weight distribution can be obtained by this method in a very short time compared with other classical methods. Therefore, this method is now being most widely utilized for solution characterization of polymers. However, the absolute molecular weight (and molecular weight distribution) cannot be measured by this method. That is why the combination of low-angle laser light scattering (LALLS) and GPC was proposed for an improved method, the advantage of which is based on the ability of LALLS to measure continuously the absolute molecular weight and also to detect minute concentrations of higher-molecular-weight species such as microgel.[1] Many papers using the combination of GPC and LALLS have been published in 1983–1986.[2–6]

Hjertberg demonstrated the effectiveness the combination of GPC and LALLS and the shortcomings of melt flow index (MFI) and density as the only characteristics of polyethylene.[1] In Fig. 1, two samples have about the same molecular weight distribution but a different degree of long chain branching which accounts for the difference in MFI. In Fig. 2, the molecular weight distributions of series A within which the MFI is approximately constant are shown. The distributions are very different and these differences are still more pronounced for the high-molecular-weight species as determined by LALLS. In some cases, minor quantities of extremely high-molecular-weight fractions will influence the rheological properties. The GPC LALLS technique offers a very effective method.[1] In Fig. 3, film and coating qualities of low-density polyethylene are compared. The refractive index (RI) traces are similar. The light-scattering traces of both qualities are bimodal with a dominating peak at an elution volume where

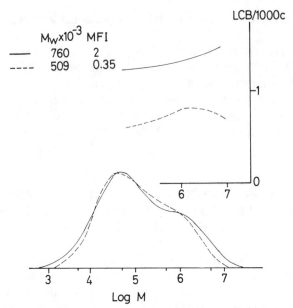

FIG. 1. Molecular weight (*M*) and long chain branching (LCB) distributions for one sample for each series as determined by the combination of on-line LALLS and viscometry: ----, from series A; ———, from series B. Within each series, melt flow index (MFI) is approximately constant: 0·3–0·4 g/10 min in series A and 2 g/10 min in series B.[1]

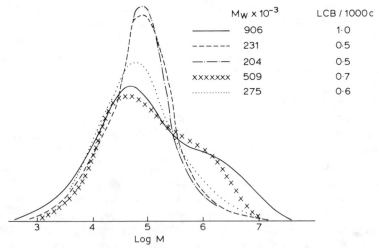

FIG. 2. Molecular weight distributions of the samples in series A determined by the LALLS detector. The value of long chain branching was obtained as the branching density at $\bar{M}_w$ with $b = 0·7$. In spite of the same melt flow index value the distributions are very different.[1]

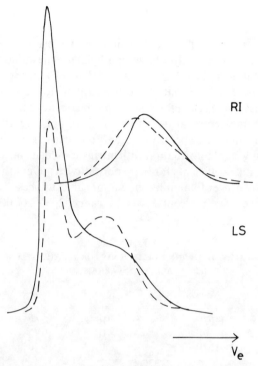

FIG. 3.   LALLS and RI traces for two different qualities of LDPE; ———, coating
quality; – – – –, film quality.[1] $V_e$, elution volume.

no signal is obtained from the RI detector. It can be estimated that the
molecular weight must be in the order of $10^8$.

New polymers recently developed are often hardly soluble or even
insoluble in conventional solvent systems. In these cases, it is very useful for
GPC measurements to find the special solvent or solvent system for these
polymers. For polymers whose solution temperatures are beyond the range
of commercial instruments (above about 150°C), the development of a new
system which allows measurement at higher temperature is required.

Many new high-performance polymers and engineering plastics have
appeared on the market in recent years and most of these polymers hardly
dissolve in conventional organic solvents at room temperature or at
relatively low temperatures. Therefore, molecular weight and rheological
characterization of these polymers remain to be solved. In conducting
molecular weight characterization studies on such typical commercial

polymers as poly(ethylene terephthalate) and polyamides, many difficulties have existed.

Berkowitz[5] reported for the first time the viscosity–molecular weight characterization of poly(ethylene terephthalate) in hexafluoroisopropanol (HFIP), pentafluorophenol (PFP), and HFIP/PFP using GPC/LALLS measurements. These solvents have been found to be superior and are capable of dissolving poly(ethylene terephthalate) at room temperature or at a relatively low temperature with no signs of polymer degradation and the measurements were therefore successfully carried out. By combining the molecular weight information with intrinsic viscosity measurements the Mark–Houwink coefficients for these poly(ethylene terephthalate)–solvent systems have been evaluated as shown in Table 1. These coefficients are applicable over the molecular weight range 2000–200 000. Kinugawa

TABLE 1

MARK–HOUWINK COEFFICIENTS FOR POLY(ETHYLENE
TEREPHTHALATE) IN VARIOUS SOLVENTS[5]

| Solvent | K | a |
|---------|-----|-----|
| HFIP | $5.20 \times 10^{-4}$ | 0.695 |
| PFP | $3.85 \times 10^{-4}$ | 0.723 |
| HFIP/PFP | $4.50 \times 10^{-4}$ | 0.705 |

The Mark–Houwink equation: $[\eta] = KM_v^a$.

reported detailed conditions of the GPC/LALLS measurements for routine work in the poly(ethylene terephthalate)–HFIP system.[7] He subsequently reported that HFIP could be applied as the solvent for GPC/LALLS measurements of various aliphatic polyamides at room temperature.[8]

Devaux et al. studied the molecular weight determination of a poly(aryletherether ketone) (PEEK), which is one of the high-temperature-resistant engineering polymers, by means of light scattering, solution viscosity and GPC.[9] GPC analyses at 115°C were performed in a 50/50 phenol–trichlorobenzene solution which provided a useful method for the determination of the molecular weight and molecular weight distribution of PEEK.

Poly(phenylene sulfide) (PPS) is another high-temperature-resistant engineering polymer and has become important because of its very excellent characteristics as an engineering plastic. Since the solution

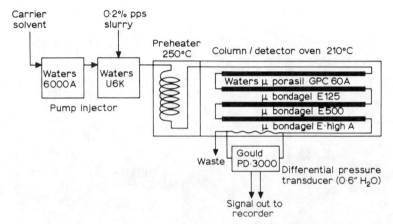

FIG. 4.   A block diagram of the instrument for a poly(phenylene sulfide) GPC system.[10]

temperature of PPS is beyond the range of commercial instruments, which at best extend to 150°C, special considerations are required to make the GPC analysis of PPS possible. Stacy developed a method for the measurement of molecular weight distribution of PPS by high-temperature GPC.[10] A diagram of the instrument is shown in Fig. 4.[10] The experimental results for a linear sample of PPS are shown in Fig. 5[10] in comparison with

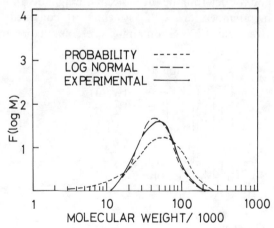

FIG. 5.   A result for a linear poly(phenylene sulfide) sample compared with theoretical molecular weight distributions.[10]

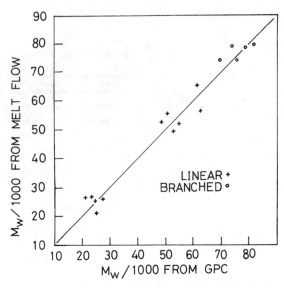

FIG. 6.   Comparison of molecular weight from GPC with molecular weight from melt flow.[10]

the usual probability distribution. Figure 6[10] shows a comparison of the molecular weight from GPC with the molecular weight from melt flow. The values are consistent within the scatter of the data. Kinugawa developed another method for determining the molecular weight distribution of PPS by a very high-temperature GPC system.[11] It consisted of a high-temperature oven held at 220°C containing a gel column for separation and

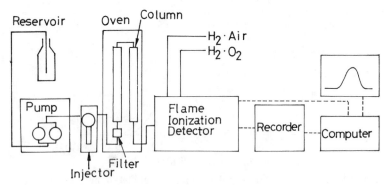

FIG. 7.   A block diagram for a very high-temperature GPC system.[11]

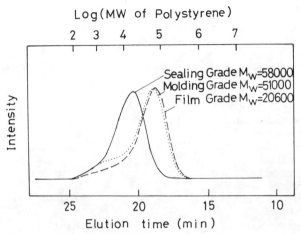

FIG. 8.  Comparison of distribution curves for three different qualities of PPS:
-----, film grade; ··········, molding grade; ———— sealing grade.[11] Column, Shodex
A80M/S(2); solvent 1-chloronaphthalene; temperature, 220°C, detector, FID,
210°C.

a flame ionization detector for detecting the concentration of PPS solution
in 1-chloronaphthalene. The diagram of the instrument is shown in Fig. 7
and the experimental results for three different qualities of PPS are shown
in Fig. 8.[11]

# 3 CHARACTERIZATION BY THERMAL TECHNIQUES

According to the definition by the International Confederation for
Thermal Analysis (ICTA), thermal analysis is defined as 'a group of
techniques in which the physical properties of a substance is measured as a
function of temperature whilst the substance is subjected to a controlled
temperature programme'. Among the thermo-analytical techniques whose
terminology was recommended by a Nomenclature Committee of ICTA
are various techniques which can be utilized for polymer characterization.
These are summarized in Table 2. In this section, the recent advances in
polymer characterization and analysis by means of these techniques will be
reviewed.

## 3.1 Differential Scanning Calorimetry (DSC)
At the 8th meeting of the ICTA, Wunderlich reported on the progress of the

TABLE 2
SUMMARY OF THERMOANALYTICAL TECHNIQUES FOR POLYMER
CHARACTERIZATION

| Technique | Abbreviation | Physical property |
|-----------|--------------|-------------------|
| Evolved gas detection | EGD | Mass |
| Evolved gas analysis | EGA | Mass |
| Differential scanning calorimetry | DSC | Enthalpy |
| Differential thermal analysis | DTA | Temperature |
| Thermodilatometry | | Dimensions |
| Thermogravimetry | TG | Mass |
| Thermomechanical analysis | TMA | Mechanical |
| Thermophotometry | | Optical |

Advanced Thermal Analysis System (ATHAS), the details of which were published in *Thermochimica Acta*.[13] This paper summarizes well his work covering the ATHAS databank, instrumentation, theory, polymer analyses, melting transition, mesophase transitions and heat-dependent properties of solid polymers. Wunderlich discussed a series of 10 polyoxides and polyolefins for which all thermal properties were known from 0 K to the beginning of decomposition in the melt, and he paid special attention to the glass transition of semicrystalline polymers which indicate a wide variety of structural sensitive effects.[13] He concluded in his review[13] not only that the equilibrium thermodynamic properties can be established and linked to fundamental, atomic-scale properties; but also that it is possible to discuss non-equilibrium states such as are found in glassy, mesophase, partially crystalline, and microphase-separated polymers. The various states can be characterized quantitatively through their heat capacity baseline and their transition behavior.

Richardson discussed detailed treatments of the quantitative aspects of three applications of DSC which have widespread usage in the polymer industry namely the glass transition, heats of reaction (i.e. curing) and crystallinity.[14] He showed that a non-critical approach could result in errors that are far beyond the instrumental uncertainties and, by contrast, correct data treatment allowed the rapid and accurate characterization of materials such as glass which have long been regarded as having little to offer but a qualitative generalization.[14] For example, Fig. 9 clearly shows the effects of thermal pretreatment on the heat capacity curves of a cured

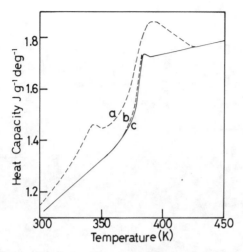

FIG. 9.  The heat capacity of a cured epoxy pipeline coating: (a) as received; (b) after cooling at 40 K min$^{-1}$ from 430 K; and (c) as (b) but cooled from 530 K. The glass transition temperature $T_g$ is about 1 K higher than after treatment (b). All heating rates are 20 K min$^{-1}$.[14]

epoxy pipeline coating by the transformation of the curve for the sample as received (a) to the comparable curves (b) and (c).[14]

Figure 10 shows the heat capacity curves for drawn PA 6/12 (nylon) filaments over the temperature range from 340 to 530 K.[14] The curves are complex and raise a number of questions, all of which could however be answered. The melting behavior of polymers is important in the fields of polymer technology, for manufacturing fibers, plastic molds and films as well as in the practical use of these products at higher temperature ranges governed by their melting characteristics. Todoki developed three types of techniques for obtaining the DSC melting curves of drawn, crystalline polymers.[15] These techniques are as follows: (1) measuring the melting curve without reorganization of imperfect crystals during the DSC scan; (2) the conventional DSC technique with scan-induced reorganization of the crystals; and (3) measuring the melting curve of the sample prevented from shrinking during the analysis. By utilizing these techniques, detailed information about the manufacturing conditions and thermal histories of unknown drawn polymers could be obtained. For example, Fig. 11 shows DSC curves of an original nylon 66 yarn and falsely twisted bulky yarns.[15] By increasing the hot-plate temperature, the melting point of the sample increases toward the equilibrium melting point of nylon 66. The shape of

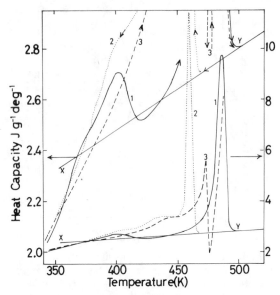

FIG. 10. The heat capacity of 6/12 nylon; 1, drawn filaments (———); 2, crystallization from the melt (⋯⋯); and 3, remelting (– – –). The fine-scale curves (left-hand ordinate) show the low-temperature thermal events in more detail. XY lines show base lines. All rates are 20 K min⁻¹.[14]

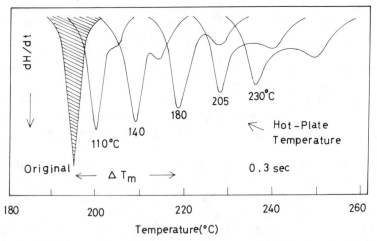

FIG. 11. DSC curves of an original nylon 66 yarn and the falsely twisted bulky yarns prepared at various hot-plate temperatures.[15]

the curve varies from unimodal in the original yarn to bimodal, indicating the formation of two types of crystals differing in their size and perfection.

### 3.2 Differential Thermal Analysis (DTA)

Wunderlich for the first time studied the crystallization behavior of polyethylene under an elevated pressure.[16] A number of studies have since been carried out on DTA, thermodilatometry, X-ray diffraction, as well as infrared and Raman spectroscopy under elevated pressures. Maeda et al.[17] developed a DTA instrument which could be used for measurements up to 700 MPa. The outline of the instrument is shown in Fig. 12. They developed a wide-angle X-ray diffraction instrument for measurements under elevated pressures while at the same time the study of the thermal properties of polyethylene was carried out by means of both instruments. Figure 13 shows the results of the thermal analyses of polyethylene, where an exothermic peak at a higher temperature range and a main exothermic peak at lower temperature range, corresponding respectively to the crystallization of hexagonal structure and the transformation of hexagonal

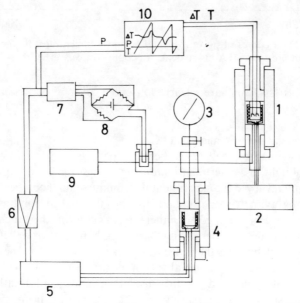

FIG. 12.   A block diagram of the high-pressure DTA instrument: 1, DTA unit; 2, temperature-controlling circuit; 3, pressure meter; 4, oil reservoir; 5, SCR controller; 6, microvoltmeter; 7, digital pressure meter; 8, Manganin gauge; 9, hand pump; 10, recorder.[17]

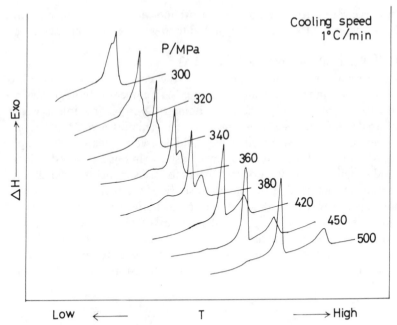

FIG. 13.   DTA curves of polyethylene at various temperatures by means of the
high-pressure DTA instrument. All cooling rates are 1°C min$^{-1}$.[17]

to rhombic systems, are clearly observed over the pressure range
500–320 MPa.

### 3.3 Thermogravimetric analysis (TGA)

Thermogravimetric analysis is being widely used for compositional
analysis, weight loss determination during processing and decomposition.
MacCallum reviewed the experimental techniques and theoretical analyses
of thermogravimetry as applied to the thermal decomposition of
polymers.[18] He highlighted the inherent difficulties in thermogravimetry
when used as a means for assessing thermal stability. This is well illustrated
by Table 3, which shows the range of values published for $E$, the apparent
activation energy for thermal decomposition of two widely used
polymers.[18] He concluded that little useful mechanistic information could
be derived simply by measuring weight loss during heating and the
apparent activation energy, determined by isothermal analysis, would
apply only within the temperature range defined by the conditions of data
collection.

TABLE 3

RANGE    OF    ACTIVATION    ENERGIES    OBTAINED
EXPERIMENTALLY[18]

| | Activation energy, $E$ ($kJ\ mol^{-1}$) |
|---|---|
| Polytetrafluoroethylene | 309 |
| | 297 |
| | 327–190 |
| | 332 |
| Polystyrene | 252 |
| | 244 |
| | 210 |

TGA coupled with mass spectrometry (MS) has been a useful tool for evolved gas analysis for over a decade and some limitations that remained in this system have been overcome by coupling the TGA to an atmospheric-pressure chemical ionization (APCI) triple quadrapole mass spectrometer.[19] A schematic diagram of this system is shown in Fig. 14. This system was used for characterizing weight loss processes associated with cure and decomposition of a phenolic resole resin.[19] Weight loss occurs in

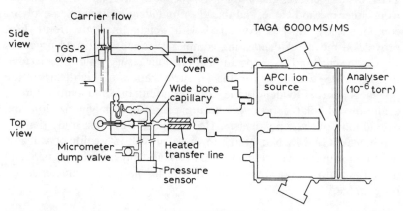

FIG. 14.    A schematic diagram of TGA/MS/MS interface between a Perkin–Elmer TGS-2 thermogravimetric analyzer and a Sciex TAGA 6000 mass spectrometer. The atmospheric-pressure chemical ionization (APCI) source of the Sciex system permits the TGA to operate under a variety of atmospheric environments and allows for easy coupling and decoupling of the two instruments.[19]

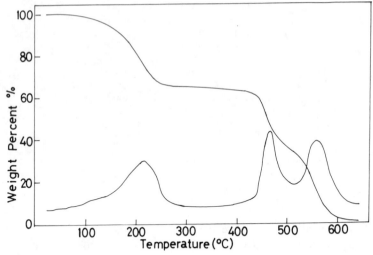

Fig. 15. Thermogravimetric analysis of phenolic resole resin, showing three weight loss steps: one during cure and two associated with decomposition. The heating rate is $10°C min^{-1}$ in air.[19]

three steps as shown in Fig. 15 and each region was examined by this TGA/MS/MS system. Prime concluded that the coupling of APCI/MS/MS to TGA vastly increases the amount of chemical information one obtains from either method alone, and that the rapid identification of co-evolving compounds by MS/MS, combined with weight loss data from TGA, provides a means of delineating complex thermally activated processes such as the cure and decomposition of phenolic resins. Goldfarb also used the TGA/MS technique for the characterization of high-temperature polymers in order to identify the nature, amount, rate and temperature of evolution of volatile products during processing, polymerization and degradation of these polymers.[20] He concluded from this study that the combined TGA/MS had been shown to provide a wealth of valuable information in the characterization of new high-temperature polymers, and it continues to play an important role in his polymer characterization programme.[20]

### 3.4 Thermomicrophotometry (TMP)
Thermomicrophotometry is the combination of thermomicroscopy and thermophotometry. TMP detects the crystalline melting point, polymorphic transformation, thermal decomposition, and glass transition

temperature. This technique is unique in that it detects changes in residual stress and polymer orientation so that isothermal studies of crystallization or stress-relaxation can be made.[21] Several studies have been reported related with the application of TMP to polymer characterization. Reffner recently described the principle and equipment of TMP and presented three examples of the application of TMP to polymer analysis and characterization.[21] Figure 16 shows polarized-light TMP and DSC data for an

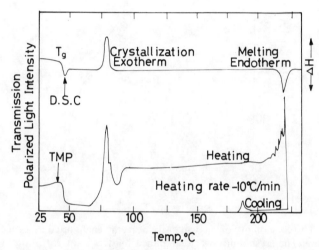

FIG. 16.   Polarized-light TMP and DSC data for a polyglycolic acid extrudate. The intensity change at 45–50°C is the glass transition, at 70°C is the onset of crystallization, and those changes above 200°C are the result of the polymer melting. The correspondence of the TMP data with those from DSC is illustrated.[21]

extrudate of polyglycolic acid, which is a biodegradable, crystalline, fiber-forming polymer.[21] The correlation of the TMP data with those from DSC is illustrated. The direct measurement of the degree of crystallinity can be made by density determination and X-ray diffraction analysis, while TMP is an effective method for recording the change in polymer crystallinity with temperature or time. Figure 17 shows X-ray and TMP data for the crystallinity of polyethylene over the temperature range 30–120°C.[21] The time required to obtain TMP data was less than one hour, which is much shorter than that taken for X-ray determination of crystallinity. These typical examples show that TMP is a thermal analytical method with a unique application to polymers and plastics.

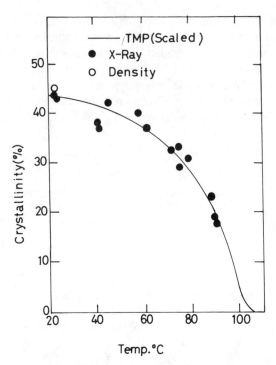

FIG. 17. The crystallinity of a low-density polyethylene plotted as a function of temperature. The crystallinity was determined by density at 25°C and by X-ray diffraction from 25 to 90°C. The polarized-light TMP intensities were scaled to the mean room-temperature crystallinity and plotted as a function of temperature.[21]

### 3.5 Thermomechanical Analysis (TMA) and Dynamic Thermomechanical Analysis (DTMA)

Thermomechanical analysis and dynamic thermomechanical analysis are effective techniques for studying the effect of not only molecular structure but also phase morphology and filler addition on the physical properties required for component design. Whilst DSC/DTA techniques give quantitative measurements of heat changes during the first-order thermodynamic transitions (e.g. melting and crystallization) and rather low resolution of the second-order transitions, the dynamic thermomechanical method detects molecular relaxations such as those occurring at the glass transition temperature with a much higher sensitivity than that achieved by DSC/DTA and hence can measure the secondary relaxations quantitatively. Gearing & Stone utilized a DMTA instrument shown

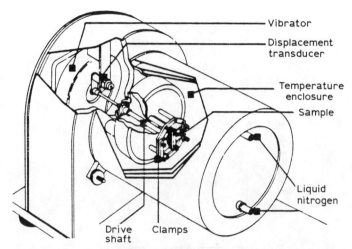

Vibrator

Displacement
transducer

Temperature
enclosure

Sample

Liquid
nitrogen

Drive    Clamps
shaft

FIG. 18.  Mechanical head of The Polymer Laboratories DMTA showing the essential features of sample mounting, vibrator system, and transducer.[22]

schematically in Fig. 18[22] to study phase-separated and non-phase-separated copolymers, rubber toughening, filler reinforcement of rubbers, carbon fiber reinforcement of epoxy resins, and frequency multiplexing for the rapid generation of relaxation spectra and activation energies.[22]

### 3.5.1  Use of DMTA in Copolymers

In random copolymers, two inherently incompatible polymer sequences are forced to co-exist within a single phase whereby the polymer exhibits a single relaxation process intermediate between those of the parent homopolymers. Figure 19 illustrates this for poly(ethylene–vinyl acetate) (random copolymer).[22] In contrast, incompatible sequences in block copolymers show phase separation. Figure 20 shows this for a block copolymer of polystyrene and random butadiene–styrene polymer.[22] Figure 21 shows an example of frequency multiplexing for an epoxy resin.[22] A complete and accurate viscoelastic characterization can be achieved by this new technique in a short period of time.

These results show that the DMTA technique provides rapid qualitative characterization of multi-phase composite polymers and frequency multiplexing is a convenient and rapid method for measuring viscoelastic response.[22] Gearing & Stone also showed that engineering data can be obtained,[22] where the theory of the mechanical response of a two-phase system was presented[22] via the Takayanagi model approach.[23]

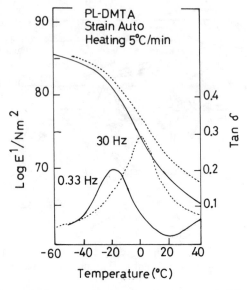

FIG. 19.  Ethylene–vinyl acetate random copolymer at two different frequencies on the PL-DMTA (PL, Polymer laboratories). A single-phase polymer exhibits a single $T_g$-relaxation between those of the parent polymers.[22]

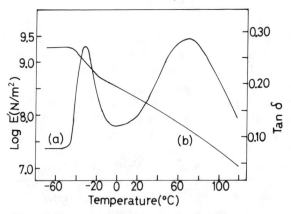

FIG. 20.  PL-DMTA scan of block copolymer of butadiene with styrene–butadiene polymer at 1 Hz and $5°C \, min^{-1}$ heating rate. The two loss peaks clearly show phase separation but with considerable phase boundary mixing.[22] (a), Rubber $T_g$ relaxation; (b), p(Styrene) $T_g$ relaxation.

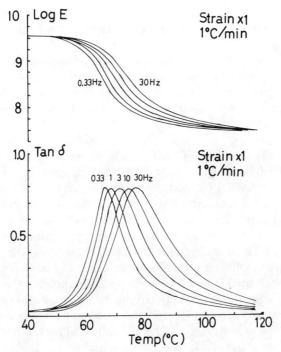

FIG. 21. Frequency multiplexing with the PL-DMTA. The sample was a moderately cross-linked epoxy resin. These data were obtained at $1°C\,min^{-1}$ giving a total measurement time of 80 min.[22]

Kaisersberger[24] studied a comparison between the performance of TMA, DMTA and DSC in order to determine the viscoelastic properties of a thermoplastic polyurethane and unidirectional carbon fiber reinforced epoxy resin. He found that the dynamic thermomechanical analysis methods were well suited to determine viscoelastic properties of solids and the sensitivity for transition processes surpassed that of the DSC technique.

### 3.6  Some other aspects

Since Du Pont introduced its first thermal analyzer in 1962, instrumentation for a number of different techniques such as DTA, DSC, TGA, TMA and DTMA have been established.[25] Up to four of these instruments may now be linked together in simultaneous operation for computer analysis by using the Du Pont 9900 Computer Thermal Analyzer. The system also incorporates a multi-tasking capability which permits data to be collected in real time while the thermal analyst performs data analysis on previously run experiments or can set up experimental conditions for future experiments.

Collaborative studies on the analysis and characterization of polymers and plastics have been recently carried out for various purposes. This type of activity is unique and must be very useful when characterizing the particular polymer samples used and assessing the relative values of the various techniques used.

### 3.6.1 Evaluation of Various Characterizing Methods

The Technical Panel PTP3 of the Technical Cooperation Programme carried out a collaborative study on the characterization of a bisphenol A epoxy resin for assessing both the relative values of the techniques for analysis and characterization and the degree of characterization which is necessary to ensure consistency and reliability in service.[26] A variety of techniques were used to characterize samples of nominally the same epoxy resin prepared in different countries. The study showed that some of the techniques indicated a difference in resin composition but these were not reflected in differences in mechanical properties. For structural analysis and specification purposes, GPC, high-performance liquid chromatography and FT-IR spectrometry are of particular value and, for the overall study of curing phenomena, DSC is specially valuable.

### 3.6.2 Collaborative Studies

Collaborative studies for improving the techniques for analysis and characterization of polymers and plastics are being continuously developed by a cooperative research programme involving governmental and public research institutes in Japan. A three-year collaborative study on the thermal analysis of plastics has been undertaken by 26 research institutes.[27] The main purpose of this study is to find out and compare the present condition and level of each organization with respect to thermal analysis determination. Five kinds of polymer samples, namely low-density polyethylene, high-density polyethylene, polypropylene, polystyrene and poly(methyl methacrylate), were used for assessing techniques for thermal analysis such as TGA, DTA and DSC. The results of the first year's study showed that differences existed in the measured values of various transition and peak temperatures due to the differences in reading the measured curves. The use of the same techniques to characterize the samples gives information on the reliability of the measurement. Further results on the second and third year's studies will be published in the near future.

## 4 NEW DEVELOPMENTS IN POLYMER SURFACE ANALYSIS AND CHARACTERIZATION BY SPECTROSCOPIC METHODS

In many industrial fields, the science and technology relating to the surface of polymers is becoming increasingly important year by year. Problems related to surfaces and interfaces are very often important when considering the particular uses for polymeric materials. Characteristics such as surface appearance, luster, adhesiveness, wettability, lubrication, electric charging, printability, blood affinity, etc., are very often important factors which dominate product's characteristics.

The technologies and instruments for the analysis and characterization of polymer surfaces have shown remarkable developments. Over the past decade, surface analytical tools and technologies with very high resolutions have been developed. Some examples are Auger electron spectroscopy (AES), secondary-ion mass spectrometry (SIMS) and X-ray photoelectron spectrometry (XPS or ESCA). A number of unique developments related to the analysis and characterization of surfaces and interfaces have been achieved by means of infrared and Raman spectroscopy.

### 4.1 X-ray Photoelectron Spectrometry (XPS)

Among surface and interface analysis techniques, XPS is easily applicable to polymeric materials and has been contributing to problem-solving in the fields of polymers and plastics during the last decade. To overcome one of the deficiencies of XPS, which is small dynamic range of core-level chemical shifts and ambiguity of binding energies, polymer surface derivatization techniques, which consist of the use of a reagent reacting selectively with a particular functional group of interest and introducing into the surface a label atom which is easily recognized in the XPS and capable of quantification, have been studied by several research groups.[28]

To enhance the adhesive characteristics of polyolefin films, which have inherently poor wettability and require a pretreatment before such operations as printing, coating, adhesive bonding or laminating, various pretreatments such as corona discharge have been studied. The derivatization techniques were applied to the pretreated olefine surfaces so as to characterize the functional groups introduced onto the polymeric surface by the pretreatments. Briggs studied various derivatization reactions to probe the functionality of the polyethylene surface that was treated by means of corona discharge.[28] The derivatization reactions employed are listed in Table 4. Full experimental details are described in Refs 29 and 30. The consensus view in the literature for the likely mechanism during

## TABLE 4
### DERIVATIZATION REACTIONS EMPLOYED[28]

$$-\underset{\underset{O}{\|}}{C}- + \left(\hspace{-2pt}\boxed{F}\hspace{-2pt}\right)\!\!-\!NH\!-\!NH_2 \longrightarrow C\!\!=\!\!N\!-\!\overset{H}{N}\!\!-\!\!\left(\hspace{-2pt}\boxed{F}\hspace{-2pt}\right)$$
(PFPH)

$$-CH_2-\underset{\underset{O}{\|}}{C}- + Br_2/H_2O \longrightarrow -CBr_2-\underset{\underset{O}{/\!/}}{C}-$$

$$-CH\!\!=\!\!\underset{\underset{OH}{|}}{C}- + ClCH_2-\underset{\underset{O}{\|}}{C}-Cl \longrightarrow -CH\!\!=\!\!\underset{\underset{O}{|}}{C}-O-\underset{\underset{O}{\|}}{C}-CH_2Cl$$
(CAC)

$$-CH_2-\underset{\underset{OH}{|}}{C}- + (acac)_2Ti(OPr)_2 \longrightarrow -CH_2-\underset{\underset{OPr'}{/}}{C}-O-Ti(acac)_2$$
(TAA)

$$-\underset{\underset{O}{\|}}{C}-OH + NaOH \longrightarrow -\underset{\underset{O}{\|}}{C}\!\!=\!\!O^-Na^+$$

$$-C-OOH + SO_2 \longrightarrow C-O-SO_2OH$$

Abbreviations: PFPH, pentafluorophenylhydrazine; CAC, chloroacetyl chloride; TAA, di-isopropoxytitanium bisacetylacetonate.

## TABLE 5
### QUANTIFICATION OF FUNCTIONAL GROUPS[28]

| Reaction | XPS ratio[a] (core level/$C_{1s}$) | Atomic ratio (element/carbon) | No. of functional groups per original surface —$CH_2$— | |
|---|---|---|---|---|
| PFPH | $(F_{1s})$ 0·205 | $5\cdot5 \times 10^{-2}$ | C=O, | $1\cdot1 \times 10^{-2}$ |
| Br$_2$/H$_2$O | $(Br_{3d})$ $3\cdot6 \times 10^{-2}$ | $10\cdot6 \times 10^{-3}$ | CH$_2$C=O, | $5\cdot3 \times 10^{-3}$ |
| CAC | $(Cl_{2p})1\cdot3 \times 10^{-2}$ | $6\cdot0 \times 10^{-3}$ | C—OH, | $6\cdot0 \times 10^{-3}$ |
| TAA | $(Ti_{2p_{3/2}})$ $6\cdot2 \times 10^{-2}$ | $1\cdot5 \times 10^{-2}$ | C—OH, | $1\cdot5 \times 10^{-2}$ |
| NaOH | $(Na_{1s})$ $8\cdot8 \times 10^2$ | $1\cdot1 \times 10^{-2}$ | —COOH, | $1\cdot1 \times 10^{-2}$ |
| SO$_2$ | $(S_{2p})7\cdot6 \times 10^{-3}$ | $4\cdot7 \times 10^{-3}$ | C—OOH, | $4\cdot7 \times 10^{-3}$ |
| None | $(O_{1s})$ 0·209 | $8\cdot7 \times 10^{-2}$ | | |

[a] Estimated error $\pm5\%$ for a given sample, $\pm15\%$ for the complete experiment.

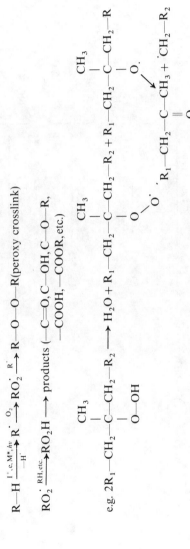

FIG. 22.  Likely mechanism of discharge treatment of low-density polyethylene.[28]

## TABLE 6
### DERIVATIZATION REACTIONS EMPLOYED[31]

| Model compound | Reaction formula |
|---|---|
| H₂N—⟨benzene⟩—O—⟨benzene⟩—NH₂ | $R-NH_2 \xrightarrow{\text{Gas-}C_6F_5CHO \text{ (PFB)}} R-N=CHC_6F_5$ |
| —(CH₂—CH)ₙ— \| COOH | $R-COOH \xrightarrow[\text{Gas-}C_6H_{11}NCNC_6H_{11}(DCC)]{\text{Gas-}CF_3CH_2OH \text{ (TFE)}} \left\{ \begin{array}{c} R-COOCH_2CF_3 \\ + \\ C_6H_{11}NHCONHC_6H_{11} \end{array} \right.$ |
| —(CH₂—CH)ₙ— \| OH | $R-OH \xrightarrow{\text{Gas-}(CF_3CO)_2O \text{ (TFAA)}} \left\{ \begin{array}{c} ROCOCF_3 \\ + \\ CF_3COOH \end{array} \right.$ |
| $\left[ \phantom{}\text{N}-H \right]$ | $\left. \text{N}-H \right] \xrightarrow{\text{Gas-}(CF_3CO)_2O \text{ (TFAA)}} \left\{ \begin{array}{c} \text{NCOCF}_3 \\ + \\ CF_3COOH \end{array} \right]$ |

PFB, pentafluorobenzaldehyde; TFE, trifluoroethanol; DCC, dicyclohexylcyanamide; TFAA, trifluoroacetic anhydride.

discharge treatment of low-density polyethylene is shown in Fig. 22. The XPS data can be quantified as described in the literature[30] to give the data in Table 5.[28]

As Briggs described in one of his papers[28] on derivatization reactions in solution phase, solvents which permeate into the polymer give rise to surface reorganization and may also extract lower-molecular-weight materials. Nakayama developed a vapor-phase derivatization technique where the derivatization reactions employed are shown in Table 6.[31] The lowest detection limits and accuracies of these techniques are shown in Table 7. The lowest detection limits are found to be ten times higher than

TABLE 7
DETECTION LIMIT AND ACCURACY[31]

| Derivatization reagent | Functional group | Detection limit (%) | Error (%) |
|---|---|---|---|
| TFAA | C—OH —NH— | 0·1–0·01 | 10 |
| TFE DCC | —COOH | 0·1–0·01 | 10 |
| PFB | —NH$_2$ | 0·1–0·01 | 10 |

Detection limit without derivatization: 1–0·1%.

those of the XPS data for untreated samples. Argon plasma treatment, which is another pretreatment to enhance the wettability and adhesive characteristics of polyolefin surfaces, by introducing mainly oxygen functional groups, was applied to a polyethylene film and the functional groups introduced were quantitatively determined by means of XPS on the surface reacted by vapor-phase derivatization.[31] The results obtained by this method are shown in Table 8, which can well demonstrate the existence and quantities of various functional groups introduced by the argon plasma treatment. This derivatization technique can be applied to other polymer and plastic surfaces. It is known that the cell-affinity of a polystyrene surface can be enhanced by introducing amines into the surface, which is carried out by means of ammonia-plasma treatment. The quantitative analysis of the functional groups introduced into the ammonia-plasma-treated polystyrene surface was performed by using the vapor-phase derivatization technique and XPS. The results obtained are shown in Table 9,[31] which demonstrates a good correlation between the

TABLE 8

QUANTIFICATION OF FUNCTIONAL GROUPS IN ARGON-PLASMA-TREATED POLYETHYLENE
FILM[31]

| $-\overset{*}{C}OOH$ | $-\overset{*}{C}OOC\diagdown$ | $\diagup\overset{*}{C}{=}O$ | $\diagdown N\diagup$ |
|:---:|:---:|:---:|:---:|
| 2% | 1% | 5% | $\overset{*}{C}{=}O$ |
| | | | 2% |
| $\overset{*}{C}{-}OH$ | $\overset{*}{C}{-}N{-}\overset{}{C}{=}O$ | $\overset{*}{C}{-}O{-}\overset{*}{C}\diagup$ | $-\overset{*}{C}H_2-$ |
| 3% | 2% | 12% | 73% |

degree of treatment, the types and amounts of the functional groups
introduced by the treatment and the cell-affinity improvement.

## 4.2 Secondary-Ion Mass Spectrometry (SIMS)

As Briggs described,[28] secondary-ion mass spectrometry has several
potential advantages over XPS for polymer surface analysis, namely greater
molecular specificity (via fingerprint spectra), greater surface sensitivity
(one to two monolayers) and the capacity to operate at a high spatial
resolution. However, there are several practical problems, namely the
expected high rate of ion-beam damage and the need for charge
neutralization. These problems were studied by Briggs[32] and the
experimental conditions which allow reproducible 'static' SIMS examination of polymer surfaces were summarized.[28] The positive-ion mass
spectra from nine pure polymer film surfaces were obtained under the given
conditions, and some prominent and characteristic peaks, which are useful
in applications, were identified.[28] Figure 23 shows one of the results, the
spectrum of poly(ethylene terephthalate),[28] the prominent and characteristic peaks of which were assigned by Briggs to the possible structures.
Briggs applied these techniques to some problem-solving analyses of
polyester film surfaces where the thickness of the region to be analyzed was
too thin for analysis by infrared spectroscopy and XPS.[28] The spectra of a
chemically modified polyester film, a surface contaminated by material
transfer between asymmetrical film surfaces in reels, and an antistatic
agent-deposited polyester film were given. One of these spectra is shown in
Fig. 24.[28] The SIMS spectrum shows that the surface is poly(ethylene
terephthalate) (PET) contaminated with hydrocarbon, resulting from the
transfer of low-molecular-weight material from the low-density polyethylene (LDPE) layer to the PET surface during contact in the reeled laminate.

TABLE 9

CORRELATION BETWEEN CELL AFFINITY AND THE AMOUNT OF FUNCTIONAL GROUPS OF AMMONIA-PLASMA-TREATED POLYSTYRENE[31]

| Plasma treatment | N/C | | | O/C | | | | | Cell affinity |
|---|---|---|---|---|---|---|---|---|---|
| | —CONH— | —NH$_2$ | —NH— / —N< | —COOH | —CONH— | —OH (>C—O—C<) | C=O | SiO$_2$ | |
| Weak | 0·02 | 0·01$_5$ | 0·03$_4$ | 0·00$_3$ | 0·02 | 0·07 | | 0·002 | + + + + |
| Strong | 0·02 | 0·01$_2$ | 0·04$_6$ | 0·00$_3$ | 0·02 | 0·08 | 0·01 | 0·00$_8$ | + + + + + |
| Blank | | | | | | | 0·01$_1$ | | + + |
| | $\uparrow$ O$_{1s}$ | $\uparrow$ GCM | | $\uparrow$ GCM | $\uparrow$ O$_{1s}$ | | $\uparrow$ C$_{1s}$, $\uparrow$ O$_{1s}$ | $\uparrow$ Si$_{2p}$ | |

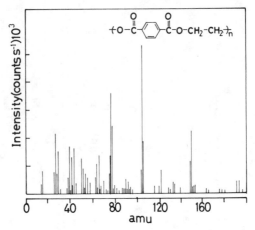

FIG. 23.    Positive SIMS of poly(ethylene terephthalate).[28]

A difficulty encountered in SIMS work on polymers is the electrical charge build-up due to very low electrical conductivities. By depositing gold film of about 500 Å thickness on the poly(ethylene terephthalate) surface, so as to avoid electrical charge build-up, very stable measurements (without using an electron gun, with good suppression of molecular ion interference and good depth resolution) can be carried out in depth profiling measurements.[33] Okuno et al. have been working extensively on

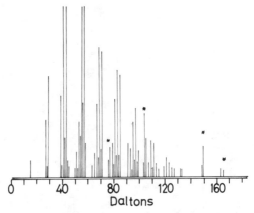

FIG. 24.    Positive SIMS of PET surface contaminated with LDPE (10³ counts s⁻¹) * Denotes peaks characteristic of PET (cf. Fig. 23).[23]

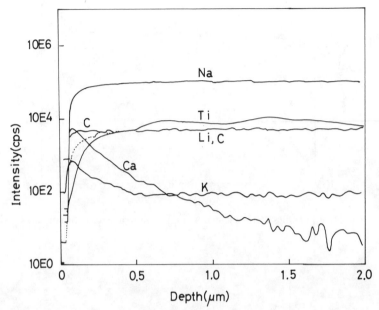

FIG. 25.   Depth profiles of lithium, sodium, potassium, calcium and titanium in a poly(ethylene terephthalate) film. The sputtered surface was confirmed by means of 'DEKTEK' (a surface roughness meter) measurement to have flatness to assure good depth resolution.[33]

depth-profiling analyses of low-concentration elements within polymers.[33] Two of the results are shown in Figs 25 and 26. Profiles shown in Fig. 25 indicate reverse in-depth variation of concentration between the Ca (calcium), K (potassium) group and the Li (lithium), Ti (titanium) group. Figure 26 shows the depth profiles of the elements in a PET/Fe–Ni/Co–Cr film (the latter two layers were prepared by sputtering metallic elements successively on to the PET surface), which indicate various diffusion characteristics depending upon each element. Another example of the depth-profiling measurement is shown in Fig. 27, which was carried out by Kurosaki,[34] on the concentration of a fluorine-containing surface-active agent in a gelatin thin layer. The concentrations of the fluorine-containing fragments derived from the surface-active agent were detected and it was found that they were at their highest near the outermost region of the layer.

### 4.3  Infrared and Raman Spectroscopy

New applications of IR and Raman spectroscopy to the surfaces, interfaces

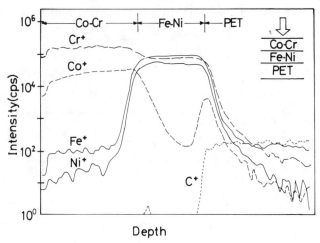

FIG. 26. Depth profiles of the elements in a poly(ethylene terephthalate)/Fe–Ni/ Co–Cr film.[33]

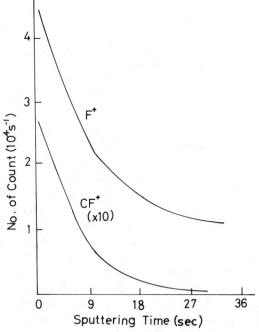

FIG. 27. Depth profiles of $F^+$ and $CF^+$ in a gelatin thin layer on a poly(ethylene terephthalate) film. These two cations come from the fluorine-containing surface-active agent blended in the gelatin.[34]

and small areas on surfaces have been developed significantly. New techniques involving IR spectroscopy include FT-IR-DRIFT (diffuse reflective infrared-FT), -ATR, -RAS, -EMS, -PAS, -SEWS and -microscopy (see Section 1). The following are some examples of applications in these fields.

Figure 28 shows a schematic illustration of the principles of FT-IR-PAS, which is very effective for the non-destructive analysis of surfaces and interfaces, and for the analysis of non-flat and black samples.[35] Figure 29

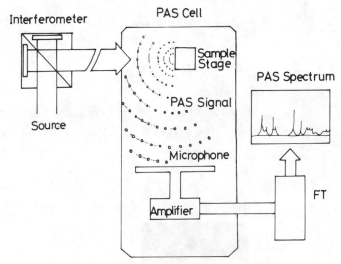

FIG. 28.    Schematic illustration of the principle of FT-IR-PAS analysis.[35]

shows one example using the FT-IR-PAS technique. When a multi-layer sample of the type illustrated in Fig. 29 was analyzed by means of FT-IR-PAS, it was possible to observe the spectrum arising from the polyurethane layer in a non-destructive manner. Thus, for example, the reaction occurring in an inner layer could be followed by means of this technique.[35]

Figure 30 is a schematic diagram depicting the new FT-IR-ATR technique. A molecular layer of cadmium eicosanoate on a glass substrate may be detected by using this particular technique as illustrated in Fig. 31.[36] Detecting an organic thin layer on a polymer substrate is influenced by the IR absorption of the substrate. Figure 32 shows the FT-IR-ATR spectra of a poly(methyl methacrylate)(PMMA) thin layer (about 130 Å) on

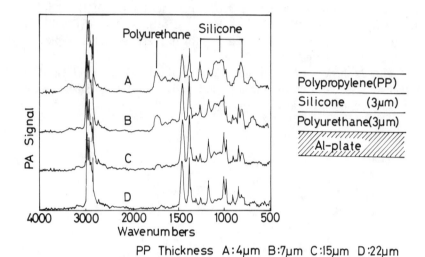

FIG. 29. An example of depth analysis possible with the FT-IR-PAS technique.[35]

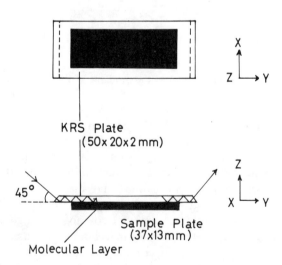

FIG. 30. Schematic diagram depicting the FT-IR-ATR technique.[36] KRS is a mixed crystal of thallium bromide and thallium iodide.

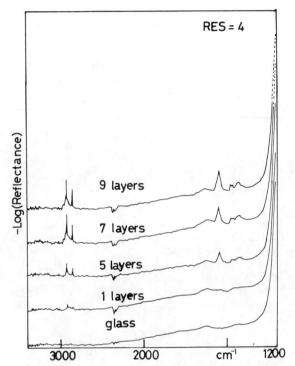

FIG. 31.    FT-IR-ATR spectra of the mono- and multi-layer films of cadmium
eicosanoate on a glass substrate.[36]

nylon 6.[37] In this case, the spectrum of PMMA can be separated by
subtracting the spectrum of nylon 6 from the measured spectrum.

Figure 33 shows a schematic illustration of the principles of FT-IR-
microscopy, which can be very effective for analyzing small inclusions,
contamination or adherents on polymer surfaces. The spectra of a thin-
sliced piece of a poly(ethylene terephthalate) fiber, whose diameter is 15 $\mu$m,
surrounded by epoxy resin are shown in Fig. 34, which clearly demonstrates
the effect of the thickness of an IR light bundle on the spectra.[35]

Increasing needs for non-destructive micro-analytical techniques
providing structural information of polymers have been stimulated
by the use of the laser Raman microprobe. These techniques can give
detailed information about the chemical composition, the crystal structure,
the intermolecular interaction and the lattice and molecular orientation on
a microscopic scale.

The degree of crystallinity of poly(ethylene terephthalate) can be

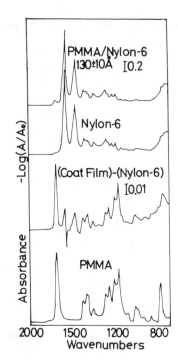

FIG. 32. FT-IR-ATR spectra of a poly(methyl methacrylate) (PMMA) thin layer (about 130 Å) on nylon 6. The spectrum of PMMA can be separated by subtracting the spectrum of nylon 6 from the measured spectrum.[37]

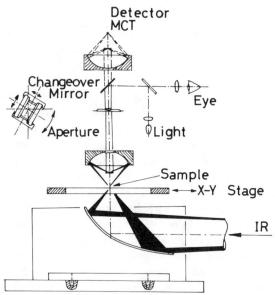

FIG. 33. Schematic illustration of the principle of FT-IR-microscopy.[35] MCT means mercury-cadmium-tellurium.

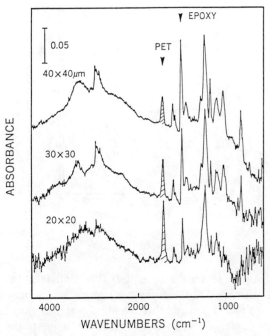

FIG. 34.   Spectra of a thin sliced piece of a poly(ethylene terephthalate) fiber, whose diameter is 15 μm, surrounded by epoxy resin.[35]

estimated by measuring the bandwidth of the carbonyl stretching vibration at 1730 cm$^{-1}$. Figure 35 shows the density profiles of the longitudinal cross-sections of uniaxially oriented PET films measured by scanning the laser beam at 10–15 μm intervals.[38] The density profiles obtained can be interpreted by the thermal history of PET films during the drawing process. As an example of the detection of an inclusion particle in a plastic film, Ishida showed the result of a $CaCO_3$ particle in a polypropylene film.[38]

The discovery of the surface-enhanced Raman scattering (SERS) effect is another attractive development as it can be applied to surface analysis. Figure 36 indicates that the Raman spectrum of a surface can be considerably enhanced and hence measured to a very high degree of sensitivity by means of silver-island film deposition.[35,38] Figure 37 illustrates the spectral changes in highly oriented pyrolytic graphite (HOPG) brought about by argon-ion etching.[35,38] A broad Raman band (1350 cm$^{-1}$) arising from the disordered carbon structure appears in the spectrum. When a silver-island film is placed upon the etched HOPG, the

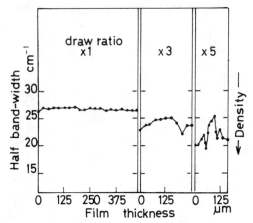

FIG. 35.　Crystallinity profiles of the longitudinal cross-sections of uniaxially oriented PET films.[39]

resulting Raman spectrum is remarkably different. In this case the outermost disordered surface has been enhanced exclusively by contact with the silver-island film and this enhanced spectrum may be measured to a very high degree of sensitivity. The result suggests the possibility of utilizing the SERS effect as a new high-sensitive surface probe.

The Raman spectrum of a very thin outermost layer can be measured by

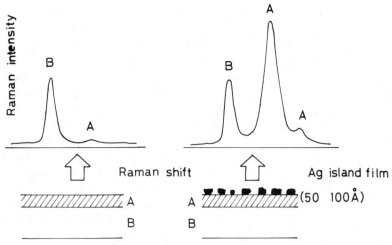

FIG. 36.　Application of SERS to surface analysis.[38]

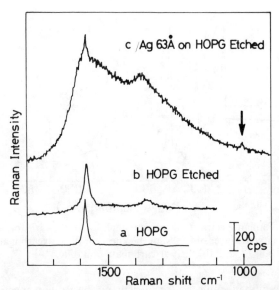

FIG. 37.   Enhanced Raman scattering arising from an argon-ion-etched HOPG surface covered with a silver-island film.[39]

ATR Raman spectroscopy. Iwamoto demonstrated that the Raman spectrum of a polymer surface can be measured by utilizing a technique whose schematic diagram is shown in Fig. 38.[39] The resulting Raman spectra of a polystyrene thin layer (60 Å) on a polycarbonate substrate are shown in Fig. 39.[39]

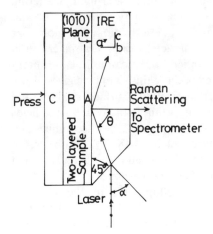

FIG. 38.   A schematic diagram of ATR Raman spectroscopy.[39] IRE means Internal Reflection Element.

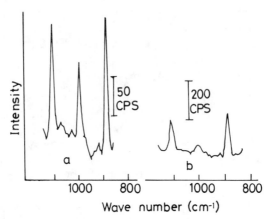

FIG. 39. The Raman spectra at the incident angle of $\theta = \theta_c + 3.7°$ for (a) the sample having a polystyrene layer of $0.006 \mu m$ thickness on the polycarbonate base layer and (b) the polycarbonate film only.[39]

The wave-guide techniques for measuring the Raman spectra of a very thin layer developed by Robolt[40] and Scherer[41] are being watched with interest because of the potential application to polymer surface analysis and structural analysis of LB (Langmuir–Blodgett) membranes.

## 5 CONCLUSION

The gel permeation chromatography/low-angle laser-light scattering (GPC/LALLS) technique, which is a combination of two most effective and instrumentally developed methods, is the key technology for molecular characterization by means of solution characterization. Characterization by thermal techniques, especially differential scanning calorimetry (DSC), has been remarkably developed because this technique gives plenty of information about polymer characterization. Surface analysis and characterization have recently become important in many industrial fields related to polymers and plastics and, for this purpose, new techniques and applications of spectroscopic methods such as X-ray photoelectron spectroscopy (XPS), secondary-ion mass spectrometry (SIMS) and infrared and Raman spectroscopy have been especially developed.

In this chapter, some important developments in the methods summarized above that have been published since 1983 have been reviewed. I hope that this chapter could become a key to open the field of analysis and characterization of polymers and plastics for readers. Other tools and techniques will be reviewed sometime in the future.

## ACKNOWLEDGEMENT

The author would like to thank Dr H. Ishida and Dr M. Todoki, both members of Toray Research Center Inc., for helpful suggestions on finding references. Toray Research Center Inc. and Toray Techno Co. Ltd, which are sister companies, are specialists in analysis and characterization, and in environmental analysis, respectively, as one of their business functions.

## REFERENCES

1. HJERTBERG, T., KULIN, L. I. & SORVIK, E., Polymer Testing, 3 (1983) 267.
2. SARTO, L. G. Jr, Soc. Plast. Eng. Annual Technical Conference, 41 (1983) 554.
3. MALIHI, F. B., KUO, C-Y. & PROVDER, T., J. Appl. Polym. Sci., 29(3) (1984) 925.
4. CAEL, J. J., CIETEK, D. J. & KOLPAK, F. J., J. Appl. Polym. Sci., Appl. Polym. Symp., 37, Pt 1 (1983) 509.
5. BERKOWITZ, S., J. Appl. Polym. Sci., 29(12), Pt 2 (1984) 4353.
6. SHIGA, S., J. Soc. Rubber Ind., Japan, 59(3) (1986) 162.
7. KINUGAWA, A., Prepr. 50th Annual Meeting Chem. Soc. Japan (April 1985) 624.
8. KINUGAWA, A. & ISHIMURO, Y., Prepr. 51st Annual Meeting Chem. Soc. Japan (April 1986) 547.
9. DEVAUX, J., DELIMOY, D., DAOUST, D., LEGRAS, R., MERCIER, J. P., STRAZIELLE, C. & NIELD, E., Polymer, 26 (1985) 1994.
10. STACY, C. J., J. Appl. Polym. Sci., 32 (1986) 3959.
11. KINUGAWA, A., Kobunshi Ronbunshu (J. Soc. Polym. Sci., Japan), 44 (1987) 139.
12. LOMBARDI, G., For Better Thermal Analysis, 2nd edn, ICTA Report, 1980.
13. WUNDERLICH, B., Thermochimica Acta, 92 (1985) 15.
14. RICHARDSON, M. J., Polymer Testing, 4 (1984) 101.
15. TODOKI, M., Thermochimica Acta, 93 (1985) 147.
16. DAVIDSON, T. & WUNDERLICH, B., J. Polym. Sci., A-2, 7 (1969) 377.
17. MAEDA, Y., KANETSUNA, H., IGUCHI, M. & KATO, M., Technical Bulletin of the Society for The Promotion of Industry and Technology in Japan, 152 (1985) 7.
18. MACCALLUM, J. R., Thermochimica Acta, 96 (1985) 275.
19. PRIME, R. B., Polymer Preprints, 26(1) (1985) 15.
20. GOLDFARB, I. J., Polymer Preprints, 26(1) (1985) 17.
21. REFFNER, J. A., American Laboratory, 16(4) (1984) 29.
22. GEARING, J. W. E. & STONE, M. R., Polymer Composites, 5(4) (1984) 312.
23. TAKAYANAGI, M., HARIMA, H. & IWATA, Y., Mem. Faculty of Engineering, Kyushu Univ., 23 (1963) 1.
24. KAISERSBERGER, E., Thermochimica Acta, 93 (1985) 291.

25. ANON., *Engineering Materials and Design*, **29**(7) (1985) 41.
26. WRIGHT, W. W., *Brit. Polym. J.*, **15** (1983) 224.
27. SERIZAWA, M., TAKAHASHI, T., *Kagaku to Kogyo* (*Chemistry and Chemical Industry*), **60**(1) (1986) 17.
28. BRIGGS, D., *Polymer*, **25**(10) (1984) 1379.
29. BRIGGS, D. & KENDALL, C. R., *Polymer*, **20** (1979) 1053.
30. BRIGGS, D. & KENDALL, C. R., *Int. J. Adhesion Adhesives*, **2** (1982) 13.
31. NAKAYAMA, Y., TAKAHAGI, T., SOEDA, F., HATADA, K., NAGAOKA, S., SUZUKI, J. & ISHITANI, A. J. Polymer Science Part A. *Polymer Chem.*, **26** (1988) 559.
32. BRIGGS, D. & WOOTTON, A. B., *Surface Interface Anal.* **4** (1982) 109.
33. OKUNO, K., TOMITA, S. & ISHITANI, A., *SIMS IV, Proc. 4th Int. Conf., Osaka*, Springer Series in Chemical Physics, Vol. 36. Springer-Verlag, Berlin, Heidelberg, West Germany, 1984, p. 392.
34. KUROSAKI, K., *Kobunshi* (*High Polymers*), **32**(9) (1983) 642.
35. ISHIDA, H. & ISHITANI, A., *Prepr. Seminar Kansai Branch Soc. Fiber Sci. Technol., Japan*, Tokyo, Japan, (Feb. 1985) p. 1.
36. OHNISHI, T., ISHITANI, A., ISHIDA, H., YAMAMOTO, N. & TSUBOMURA, H., *J. Phys. Chem.*, **82** (1978) 1989.
37. NAGASAWA, Y., to be published.
38. ISHIDA, H. & ISHITANI, A., *Proc. 9th Int. Conf. Raman Spectroscopy* (27 August–1 September 1984) p. 74.
39. IWAMOTO, R., MIYA, M., OHTA, K. & MIMA, S., *J. Chem. Phys.*, **74** (1981) 4780.
40. ROBOLT, J. F., SCHLOTTER, N. E. & SWALEN, J. D., *J. Phys. Chem.*, **85** (1981) 4141.
41. SCHERER, J. R. & BAILEY, G. F., *J. Membrane Sci.*, **13** (1983) 43.

# Chapter 3

# MODIFICATION OF PVC WITH NBR

P. W. MILNER

*Goodyear Chemicals Europe, Les Ulis, France*

## 1 INTRODUCTION

The growth in the use of polyvinyl chloride (PVC) has been phenomenal since Semon[1,2] demonstrated that PVC could be processed and converted into a product having rubbery characteristics by mixing it with high-boiling-point esters such as tritolyl phosphate and fluxing the resultant blend by heating. Before this, however, the rigid polymer, which was first reported by Baumann[3] in 1872, had not been commercially exploited to any great extent, though work had been carried out by several companies to produce an internally plasticised resin based on vinyl acetate as a comonomer[4-6] in order to facilitate processing.

Following the discovery of the action of ester plasticisers, the outbreak of the Second World War gave the required impetus to industry, which found out that plasticised PVC could be used to replace vulcanised rubber in certain applications, particularly in electrical insulations where the added advantage of its flame resistance made it the ideal product for military use. It was only after the war that the development of new products for the consumer market grew and the industry really mushroomed.

It is estimated that 85% of the PVC usage is in the plasticised form and the factors responsible for its rapid growth are considered to be: (a) low cost; (b) the ability to be compounded to meet a wide range of final application properties (both solid and blown); (c) good chemical resistance and weathering properties; and (d) the ability to be processed by a wide variety of techniques (including calendering, extrusion, injection, blow and compression moulding, coating and impregnation using latices and solutions).

The ability of PVC to be compounded with plasticisers and fillers makes it unique amongst thermoplastics in that the initial properties of the resin can be changed to give compounds suitable for a wide variety of final products. It is this characteristic that has led to the use of acrylonitrile–butadiene copolymers (nitrile rubbers) as modifiers for PVC. It has been reported[7] that as early as 1938 efforts were being made in Germany to market nitrile rubber (NBR) as a plasticiser for PVC.

This chapter reviews the modification of PVC with NBR from both the theoretical and the practical points of view.

## 2 MODIFICATION OF PVC[7-11]

Rigid PVC compounds have second-order transition points ($T_g$) above the normal temperature at which the products are used and may therefore undergo brittle fracture when subjected to impact. This has limited their use in certain critical applications and resulted in the development of a wide range of impact improvers.

### 2.1 Use of Impact Modifiers for Rigid Compounds

Many of these impact modifiers have a glass transition temperature below room temperature, i.e. they are rubbery, and when added to the PVC compound in concentrations up to 10% can increase the impact strength 20- to 30-fold. Impact modifiers are available with varying compositions but acrylonitrile–butadiene–styrene (ABS) and methyl methacrylate–styrene–butadiene (MBS) terpolymers are the most widely used. Acrylonitrile–butadiene (NBR) copolymers have not as yet found wide use as modifiers for rigid PVC, though experimentation is ongoing.

### 2.2 Use of Rubber Modifiers for Flexible Compounds

It is, however, with flexible PVC compositions that acrylonitrile–butadiene (NBR) copolymers have found increasing utilisation as property modifiers. They have become even more commercially and technically attractive due to their introduction in powder form, which allows them to be processed in low-energy, high-speed dry blenders rather than in the costly high-energy consuming internal mixers needed for mixing conventional bale rubber (see Section 7.1).

Whereas rigid PVC has only one major defect, i.e. impact resistance, flexible PVC, through its need to use liquid plasticisers, is limited in its final application usefulness due to:

(a) hardening with time—due to loss of plasticiser either by volatilisation and migration or by extraction due to contact with oils and solvents;
(b) inadequate low-temperature flexibility during service—due to conditions listed in (a);
(c) poor flex, cut growth and tear resistance;
(d) low abrasion resistance;
(e) marginal wet friction (grip) characteristics.

All these property shortcomings can be improved by incorporating NBR polymers and, in addition, certain 'rubbery' characteristics are imparted to the PVC such as lower compression set and higher elongation at break.

The success of NBR as a PVC modifier is due not only to its availability as a powder but also to its chemical composition, which confers on the polymer the required degree of compatibility necessary for effective property modification (see Section 4).

## 3 NITRILE RUBBER (NBR) TYPES AND PRODUCTION

Nitrile rubbers are available both with varying acrylonitrile contents and at varying plasticising levels. The acrylonitrile content has an influence on the degree of compatibility of the polymer with PVC and the polymer plasticity or viscosity modifies the processing behaviour of the melt during the fabrication stages.

### 3.1 NBR Types

From investigations into compatibility[12] it has been established that polymers with acrylonitrile contents between 30 and 40% of the molecular hydrocarbon content are most suitable for PVC modification.

To decide the viscosity level, consideration has to be given to the type of final processing to be used to make the product. Moulding requires a low-viscosity rubber that will have minimum effect on the melt flow during the injection whereas extrusion and calendering require some degree of dimensional stability in order to obtain a smooth high-quality surface coupled with accurate product dimensions. This can be achieved by having a polymer with a slightly crosslinked molecular structure (see Section 8).

Modern plastics mixing techniques using high-speed dry blenders make it imperative that the polymer be available in powder form (see Section 7).

## 3.2 Production

The suppliers of nitrile rubber have carried out considerable experimentation on the methods to produce powder. These vary from several mechanical grinding techniques, including cryogenic grinding, through spray and flash drying to particle encapsulation.[13,14] All methods use various amounts and types of partitioning agents to prevent 'particle re-agglomeration or blocking' of the powder during storage. Between 3 and 15 phr of inorganic materials such as talc, calcium carbonate and silica, or of organic materials such as PVC, calcium stearate or zinc stearate, are the most popular.

The only two methods being used commercially, however, are grinding and spray drying, and the relative merits of powders prepared by these different technologies revolve around (a) the non-rubber constituents, (b) the polymer viscosity, and (c) the particle size.

### 3.2.1 The Non-rubber Constituents

The non-rubber constituents are the products that are part of the emulsion polymerisation process used to prepare the polymer. These are soaps, activators and catalyst residues, and can have an effect on the clarity and the water sensitivity, and hence the electrical properties of the finished PVC blend. For the grinding process, the polymer is coagulated and washed prior to the final process and therefore the non-rubber content is reduced to a minimum. However, for spray drying, the polymer latex is fed directly through a nozzle into a heated chamber where the water is evaporated, leaving all the non-rubber constituents present in the dried polymer particle. The high level of soap remaining minimises the tendency for the powder to block in the dryer before a conventional partitioning agent can be added. In difficult cases antiblocking agents can be added to the latex prior to drying.

### 3.2.2 Polymer Viscosity

Polymer viscosity also has an influence on the blocking characteristics of the powder. The grinding process can produce powders with low viscosities more successfully than the spray drying technique. This is because the partitioning agent can be introduced to the powder more effectively during the grinding technique than it can during spray drying. Polymers with Mooney viscosities below $30 ML_4$ are particularly difficult to handle and

## TABLE 1
### NBR POWDER PARTICLE SIZE DISTRIBUTION[15]

| Particle size range (mm) | Percentage by weight |
|---|---|
| 0·0–0·075 | 1·7 |
| 0·075–0·15 | 4·0 |
| 0·15–0·30 | 16·5 |
| 0·30–0·70 | 77·8 |
| 0·70–1·00 | 0 |

## TABLE 2
### COMMERCIALLY AVAILABLE NBR POWDERS

| Trade name and number | | Nominal acrylonitrile content (%) | Nominal Mooney viscosity, $ML_{1+4}$ (100°C) | Manufacturer |
|---|---|---|---|---|
| Perbunan | N2807 NS | 28 | 45 | Bayer AG, Geschäftsbereich |
| | N3307 NS | 34 | 45 | KA, D-5090 Leverkusen, |
| | N3310 | 34 | 65 | FRG |
| | N3807 NS | 39 | 45 | |
| Hycar | 1452 P-50 | 33 | 50 | B. F. Goodrich Chemical |
| | 1422 × 8[a] | 33 | 67 | Group, 6100 Oak Tree |
| | 1422[a] | 33 | 80 | Boulevard, Cleveland, |
| | 1492 P-80 | 33 | 80 | OH 44131, USA |
| | 1411 | 41 | 115 | |
| Chemigum | P8-D[a] | 33 | 80 | Goodyear Chemicals |
| | P8B-A[a] | 33 | 80 | Europe, Avenue des |
| | P83[a] | 33 | 50 | Tropiques, ZA de |
| | PFC[a] | 33 | 50 | Courtaboeuf, 91941 Les |
| | P612-A | 33 | 25 | Ulis Cedex, France [and] |
| | P615-D | 33 | 50 | Goodyear Tire & Rubber |
| | P7-D | 33 | 86 | Co., Akron, Ohio 44316, USA |
| Nipol | DN-223 | 33 | 35 | Nippon Zeon, Furukawa Sogo Building, 6-1 Marunouchi, 2-chome Chiyoda-ku, Tokyo, Japan |
| Krynac | 1403 H176 | 29 | 50 | Polysar International SA, |
| | 1402 H24F | 33 | 50 | Route de Beaumont 10, |
| | 1402 H83 | 33 | 50 | PO Box 1063, |
| | 1402 H82 | 33 | 70 | CH-1701 Fribourg, |
| | 1122 P[a] | 33 | 50 | Switzerland |
| | 1411[a] | 40 | 115 | |

[a] Pre-crosslinked grade.

present severe problems for the spray drying process as the temperature of drying makes the polymer extremely 'tacky', and even the residual soap present in the latex cannot overcome this.

### 3.2.3 Particle Size

The particle size range that makes an NBR powder suitable for efficient dry blending with PVC has been empirically investigated by the NBR powder producers, who initially were providing crumb rubber to the adhesives industry at up to 6 mm particle diameter. The grinding process was improved, spray drying was introduced and 1 mm maximum particle diameter is now the accepted industry standard for a powder. Spray drying gives average particle sizes well below 1 mm and in certain cases gives a problem of 'fines' during processing.

Table 1 lists the particle size distribution of a commercially available powder prepared by grinding.

Due to the increasing interest in NBR powders for PVC modification the range of commercially available products is continually changing, but Table 2 lists the major producers and types available at the beginning of 1987.

## 4 COMPATIBILITY

In practice, truly compatible blends of polymers are rare; the criteria by which their compatibility is judged are diverse, and depend to a large extent on the final application demands of the resultant blend. In certain cases discreet phases are deemed necessary and miscibility is not required (e.g. the rubber reinforcement of polystyrene[16]). PVC, however, by virtue of the weakly acidic or proton-donating hydrogen, has been shown to exhibit miscibility with a considerable number of polymers,[17-19] of which acrylonitrile–butadiene copolymers (NBRs) have proved to be amongst the most widely used.

The NBRs available for PVC modification are random copolymers produced by emulsion polymerisation and having the general molecular structure

$$\left[CH_2-CH{=}CH-CH_2\right]_m \left[CH_2-\underset{\underset{CN}{|}}{CH}\right]_n$$

and it is the polarity of the nitrile grouping (—CN) that makes these copolymers compatible with PVC and thereby extremely effective as property modifiers.

Various methods have been used to determine the compatibility of polymer blends, each having its advantages and sensitivity. Microscopic methods using transmission electron microscopy (TEM)[20] and scanning electron microscopy (SEM),[21] glass transition temperature studies using thermal analysis tests,[22] dynamic mechanical and dielectric testing,[23] nuclear magnetic resonance (NMR),[24] Fourier transform infrared spectroscopy (FTIR)[25] and stress–strain[26] measurements have all been used. Of the methods listed, microscopy, glass transition temperature measurements and dynamic mechanical testing have been most valuable in studies of polymer blend compatibility. However, it should be noted that a given system which appears compatible by one method may be incompatible by another.[27]

The compatibility of PVC/NBR blend systems have been examined in detail by numerous workers[28,29] using the methods listed above. However,

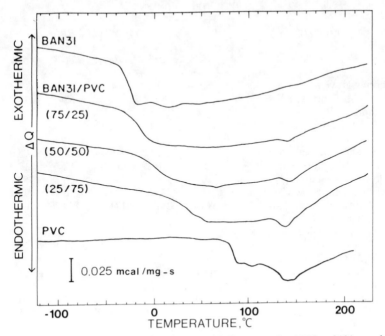

FIG. 1.   Differential scanning calorimetry (DSC) curves for NBR—31% acrylo-nitrile content (BAN 31) PVC blends. BAN, butadiene–acrylonitrile polymer.

TABLE 3
TEST COMPOUND FOR PVC/NBR ALLOY PHYSICAL PROPERTY STUDY[12]

| Formulation | (parts by weight) |
|---|---|
| PVC resin ($K$ value at 69) | 100·0 |
| DOP (dioctyl phthalate) | 65·0 |
| Stabiliser, barium/cadmium/zinc (Ba/Cd/Zn) type | 3·0 |
| Chelator, organic phosphite | 1·0 |
| Calcium stearate | 1·0 |
| Titanium dioxide, rutile | 3·0 |
| Filler, calcium carbonate ($CaCO_3$) | 15·0 |
| NBR polymer | 30·0 |

TABLE 4
PHYSICAL PROPERTY VARIATION IN PVC/NBR ALLOYS[12] AS A FUNCTION OF
ACRYLONITRILE CONTENT

| Property | Acrylonitrile in NBR[a] | | | | |
|---|---|---|---|---|---|
| | 19% | 25% | 34% | 40% | 50% |
| Specific gravity | 1·24 | 1·24 | 1·25 | 1·26 | 1·26 |
| Hardness, Shore A | 72 | 67 | 67 | 68 | 75 |
| Modulus (MPa) | | | | | |
| At 50% elongation | 2·5 | 2·6 | 2·7 | 2·6 | 3·1 |
| At 100% elongation | 3·7 | 4·0 | 4·2 | 4·2 | 5·0 |
| At 200% elongation | 5·5 | 6·2 | 6·4 | 6·5 | 8·0 |
| Tensile strength (MPa) | 5·6 | 10·2 | 12·6 | 12·4 | 12·4 |
| Ultimate elongation (%) | 220 | 370 | 470 | 460 | 350 |
| Tear resistance, die 'C' ($kN m^{-1}$) | 21·5 | 37·6 | 39·6 | 40·3 | 39·2 |
| DIN abrasion ($mm^3$ loss) | 398 | 231 | 171 | 159 | 206 |
| Ross flex | | | | | |
| At $-20°C$ (kilocycles to 500% cut growth) | 0 | 3·8 | 6·0 | 1·1 | 0 |
| At 23°C at 250 kilocycles (% cut growth) | 500 | 300 | 62·5 | 75 | 125 |
| Compression set (%) | | | | | |
| 22 h at 20°C | 27·0 | 20·0 | 20·5 | 21·0 | 31·0 |
| 22 h at 70°C | 67·0 | 67·0 | 60·0 | 68·5 | 68·5 |

[a] NBR grades used: Krynacs 1965, 2565, 3460, 4065, 5075.

it should be stressed that complete homogeneity is not always needed in order to obtain the desired property modifications.

## 4.1 Microscopy

The use of electron microscopy has established that providing an adequate fluxing temperature (see Section 7) is used, NBRs with acrylonitrile contents of 32% and above do not exhibit microheterogeneity and complete compatibility is achieved.[12,29] However, these results were obtained using linear polymers. Pre-crosslinked NBR polymers have advantages in other areas, such as remaining as distinct domains, the size of

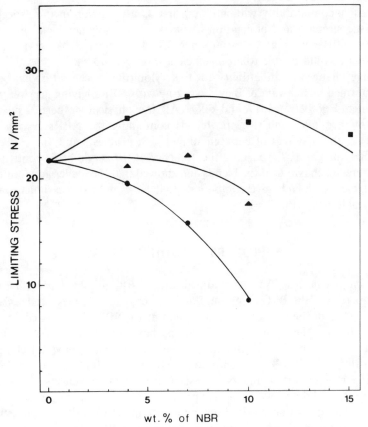

FIG. 2.    Limiting stress versus NBR content in the PVC/NBR blends for 10-s creep time:[30] ●, PVC/NBR, 21% ACN (Hycar 1024); ▲, PVC/NBR, 30% ACN (Hycar 1043); ■, PVC/NBR, 42% ACN (Hycar 1041).

which is dependent upon the shear rate of mixing and the viscosity of the NBR polymer and the resultant compound.[12]

### 4.2 Glass Transition Temperature

Compatible systems will exhibit a merging or 'smearing' of $T_g$ values situated somewhere between those of the constituent polymers,[29] whereas an incompatible blend will show the distinct $T_g$ values of the individual polymers. Figure 1 illustrates this phenomenon for a linear NBR of 31% acrylonitrile (ACN) content.[28]

### 4.3 Mechanical Testing

Gross incompatibility will show up first as a reduction in the expected tensile strength and elongation at break of the polymer blend and for PVC/NBR combinations using an NBR at or above 32% acrylonitrile content would be due to an insufficient fluxing temperature.

The change in compatibility due to acrylonitrile concentration has been confirmed by Schwarz & Bley[12] working with NBRs having acrylonitrile contents of between 19 and 50%. All the physical properties passed through a maximum for the blends containing the NBRs having an acrylonitrile content of between 30 and 40% (Tables 3 and 4).

Bergman et al.[30] measured the influence of the acrylonitrile content of the creep behaviour of the blends and showed that the higher acrylonitrile polymers in blends give compounds that have more stable dimensional characteristics (Fig. 2).

## 5 PROPERTY MODIFICATIONS

When plasticised PVC is modified with NBR, all the basic physical properties of the PVC compound undergo change. The magnitude of the change is dependent on the total amount of NBR that is used in the modification. The property modifications that are of most interest are those that can be considered as improvements to the basic physical characteristics of the PVC compounds and are the reasons for which the NBR is being used. In particular, it is the resistance to extraction effects of oils and solvents, and the retention of such properties as low-temperature flexibility and flexural resistance after immersion or contact with these agents, that are of prime importance when considering NBR as a PVC modifier. However, along with these sought-after property improvements, other properties also change and have to be allowed for when using NBR as a PVC modifier (e.g. hardness, tensile strength, etc.). The general trends in property modifications are the same for all types of NBRs, but the magnitude of the property change is dependent on the individual characteristics of the NBR used. The individual characteristics are (a) viscosity, which is a function of molecular weight, and (b) the linearity or pre-crosslinked nature of the polymer. Reviews of these property trends have been published by Giudici & Milner,[8] De Marco *et al.*[9] and Schwarz & Bley,[12] and are summarised in the following sections.

## 5.1  Hardness and Stress–Strain

### 5.1.1  Influence of NBR Mooney Viscosity

The hardness and stress–strain properties are only marginally altered by variations in the NBR polymer viscosity (either of the linear or pre-crosslinked type). Table 5 illustrates this for linear polymers with Mooney viscosities from 35 to 40.

TABLE 5
INFLUENCE OF NBR VISCOSITY ON PHYSICAL PROPERTIES[13]

| Test recipe | (parts per weight) |
|---|---|
| PVC resin (K value of 69) | 100·0 |
| DOP | 65·0 |
| Stabiliser, Ba/Cd/Zn type | 3·0 |
| Chelator, organic phosphite | 1·0 |
| Calcium stearate | 1·0 |
| Titanium dioxide, rutile | 3·0 |
| Filler (CaCO₃) | 15·0 |
| NBR polymer | 30·0 |

| Property | NBR[a] polymer viscosity | | | |
|---|---|---|---|---|
| | 35 | 50 | 80 | 140 |
| Hardness, Shore A | 67 | 67 | 66 | 67 |
| Modulus (MPa) | | | | |
| At 50% elongation | 2·8 | 2·9 | 2·7 | 3·7 |
| At 100% elongation | 4·4 | 4·4 | 4·2 | 5·8 |
| At 200% elongation | 6·6 | 6·9 | 6·3 | 8·6 |
| Tensile strength (MPa) | 11·9 | 12·2 | 12·1 | 14·4 |
| Ultimate elongation (%) | 430 | 415 | 450 | 400 |
| Tear resistance, die 'C' (kN m⁻¹) | 42·5 | 44·3 | 41·5 | 42·3 |

[a] ACN content, 34%. NBR grades used: Krynacs 3435, 3450, 34140. Similar results are obtained with pre-crosslinked polymers.

HARDNESS
Shore A

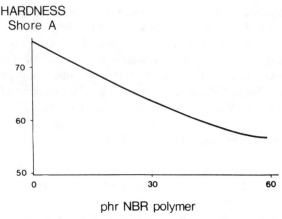

phr NBR polymer

FIG. 3.    Influence of NBR loading on hardness.

### 5.1.2 Influence of NBR Loading

Added in increasing quantities, both linear and pre-crosslinked NBRs reduce hardness and tensile strength, and increase elongation at break. The reduction in tensile strength only becomes significant above a 15-phr loading.[8]

Figures 3 and 4 illustrate those properties for a linear medium-acrylonitrile (33%) polymer of 30 Mooney viscosity (Chemigum P612) added to a PVC 100 DOP 60 base compound.

From practical evaluations it appears that if the liquid plasticiser content

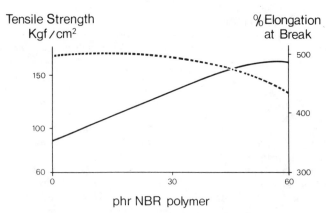

phr NBR polymer

FIG. 4.    Influence of NBR loading on tensile strength and elongation at break.

## TABLE 6
### PHYSICAL PROPERTY VARIATION AT EQUAL HARDNESS[31]

| PVC (*K* value of 70) | 100 | 100 | 100 | 100 | 100 |
|---|---|---|---|---|---|
| NBR (33% ACN, 50 ML$_4$)$^a$ | 0 | 15 | 30 | 45 | 60 |
| DIDP$^b$ | 77 | 69 | 67 | 65 | 63 |
| Ba/Cd stabiliser | 3 | 3 | 3 | 3 | 3 |
| Stearic acid | 1 | 1 | 1 | 1 | 1 |
| CaCO$_3$ | 10 | 10 | 10 | 10 | 10 |
| Hardness, Shore A | 72 | 74 | 73 | 71 | 70 |
| 100% modulus (MPa) | 8·0 | 7·6 | 6·8 | 6·1 | 5·5 |
| 300% modulus (MPa) | 14·5 | 13·6 | 12·8 | 11·8 | 11·0 |
| Tensile strength (MPa) | 15·5 | 6·5 | 16·8 | 17·0 | 17·3 |
| Elongation at break (%) | 340 | 395 | 410 | 440 | 450 |

$^a$ Chemigum P83.
$^b$ DIDP, di-isodecylphthalate.

is kept constant, there is a 1-point Shore A reduction in hardness for every four parts of added NBR.

Again from empirical work, the loss in tensile strength at loadings up to 60 phr is limited to 15% of the unmodified plasticised compound.

However, if compound modification is carried out, so that the hardness is maintained at a constant value, then the compounds become increasingly more 'elastomeric' in nature, whereby the modulus is reduced, and the elongation is increased along with an increasing tensile strength.

Table 6 illustrates the changes in stress–strain at approximately equal hardness for PVC modified by up to 60 phr NBR (33% ACN, 50 ML$_4$).

### 5.2 Abrasion Resistance

The abrasion resistance of plasticised PVC compounds is poor and this has drawbacks for its use in high-quality footwear, cables and hose covers.

The use of NBR can radically increase the abrasion resistance of a PVC compound when compounded to equivalent hardness. Figure 5 illustrates the reduction in abrasion loss that occurs when increasing quantities of NBR are added to a plasticised PVC compound with adjustments to the liquid plasticiser content in order to maintain a constant hardness.

### 5.3 Compression Set

Due to the nature of plasticised PVC, low compression set figures are not expected, but the inclusion of a pre-crosslinked NBR markedly improves the compression set characteristics. A linear polymer, however, appears to have a minimal effect.

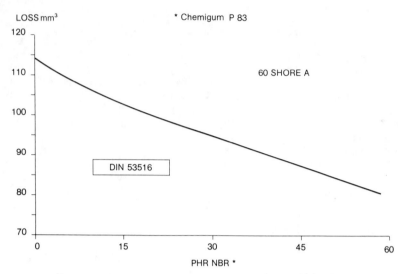

FIG. 5.   Abrasion loss (DIN 53510) of NBR-modified PVC.

Figure 6 illustrates the difference in effectiveness of pre-crosslinked linear polymers in their ability to improve the compression set.[10]

### 5.4 Flexing and Cut Growth
The flexibility and cut growth resistance during flexing of plasticised PVC compounds are moderately good at ambient temperatures. However, at lower temperatures (below 0°C) PVC compounds plasticised with monomeric plasticisers fail to meet specifications for industrial usage. The inclusion of an NBR polymer can significantly improve the cut growth resistance of PVC compounds at low temperatures. Table 7 lists the cut growth resistance of NBR-modified compounds adjusted to give equal hardness.

### 5.5 Tear Strength
The tear strength of a plasticised PVC compound is improved by the addition of NBR, and this is in line with the increased cut growth resistance already cited. Both these improvements are the result of the lower modulus inherent to compounds containing NBR. Evidence has been presented[12] that maximum tear property improvement is obtained with NBRs having acrylonitrile contents above 30%. Figure 7 illustrates this and also shows the influence of NBR loadings.

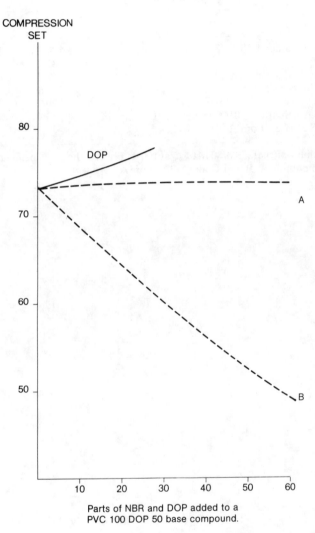

FIG. 6. Compression set: 22 h at 100°C. Polymer A, linear NBR, 33% ACN, 50 ML$_4$ (Hycar 1402 H23); polymer B, pre-crosslinked NBR, 33% ACN, 80 ML$_4$ (Hycar 1422).

TABLE 7
CUT GROWTH RESISTANCE AT $-5°C^{31}$

|  | $A^b$ | $B^b$ | $C^b$ |
|---|---|---|---|
| PVC (K value of 70) | 100 | 100 | 100 |
| DOP | 88 | 82 | 80 |
| Epoxidised soya bean oil | 5 | 5 | 5 |
| Stabilisers | 3·6 | 3·6 | 3·6 |
| NBR$^a$ | — | 20 | 30 |
| Cut growth after 150 000 cycles (mm) (BS 5131 Section 2.1) | 1·2 | 0·9 | 0·75 |

$^a$ Medium-nitrile (33%), 80 $ML_{1+4}$ at 100°C, pre-crosslinked (Chemigum P8).
$^b$ Compounds A, B and C all at 60° Shore A hardness.

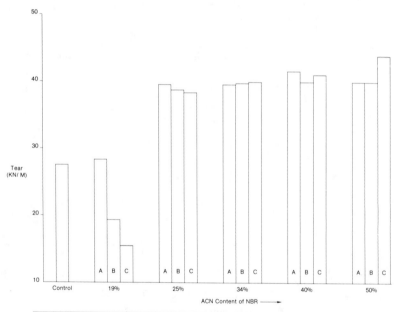

| | DOP (phr) | NBR (phr) | Hardness, Shore A |
|---|---|---|---|
| Control | 80 | 0 | 69 |
| A | 72·5 | 15 | 67–69 |
| B | 65 | 30 | 67–72 |
| C | 57·5 | 45 | 67–73 |

FIG. 7.   Tear resistance versus acrylonitrile content and NBR loading.[12] NBR grades used: Krynacs 1965, 2565, 3460, 4065, 5075.

TABLE 8
LOW-TEMPERATURE PROPERTIES[8,31]

| NBR A[a] (phr) | 0 | 10 | — | 20 | — | — | 50 | — |
|---|---|---|---|---|---|---|---|---|
| LTF[c] (°C) | −40 | −40 | — | −37 | — | — | −39 | — |
| NBR B[b] (phr) | 0 | — | 15 | — | 30 | 45 | — | 60 |
| LTF[c] (°C) | −40 | — | −37 | — | −35 | −34 | — | −33 |

[a] NBR A: 33% ACN, 80 ML$_4$ + butyl benzyl phthalate (35 phr), dioctyl adipate (15 phr) and polymeric ester (20 phr).
[b] NBR B: 33% ACN, 50 ML$_4$ + di-isodecyl phthalate (77–63 phr to give equal hardness).
[c] LTF to ASTM D-1043-72 (Clash & Berg).

## 5.6 Low-temperature Flexibility (LTF)

The improvements that can be obtained in low-temperature flexibility of plasticised PVC compounds by including NBR are marginal when compared with a plasticiser system that has been specifically designed for low-temperature work.[8] The addition of up to 50 phr of a 33% acrylonitrile NBR only improves the low-temperature stiffening temperature by 3°C (see Table 8). However, the improvement is more pronounced when the NBR is added to a standard phthalate-containing compound[31] (see Table 8).

## 5.7 Oil, Fuel and Solvent Resistance

The real value of using an NBR to improve the low-temperature properties of PVC is only fully realised when the compounds are subjected to ageing in oils, fuels and solvents. It is under these conditions that the outstanding improvements in property retention are shown. It is due to the ability of the nitrile rubber polymer not only to act as a modifier and plasticiser but also to help retain the more conventional plasticisers in the compound during the service life of the product. This retention effect is easily demonstrated by the measurement of plasticiser loss by migration and volatility (see Section 5.8).

### 5.7.1 Oil Resistance

Figure 8 illustrates the resistance to extraction of a compound containing a linear NBR, 33% ACN content at 50 ML$_4$.[10] The base compound used was:

| | |
|---|---|
| PVC (K value of 70) | 100 |
| DOP | 50 |
| Epoxidised soya bean oil | 5 |
| Ba/Cd stabiliser | 3 |
| Chelator | 1 |
| Zinc stearate | 1 |

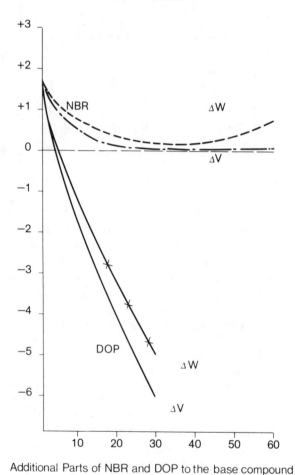

FIG. 8.   Volume and weight change in ASTM Oil No. 111 (70 h at 22°C; ASTM D471-66).

The oil resistance tests were carried out in ASTM Oil No. 111 (highly polar) at 22°C for 72 h (ASTM D471-66) and the change in volume and weight are shown. The difference between the DOP-only system and the DOP-plus-NBR system is easily seen.

### 5.7.2 Fuel Resistance

The changes in weight and volume (both wet and dry) and low-temperature

TABLE 9
WEIGHT, VOLUME AND LOW-TEMPERATURE FLEXIBILITY CHANGE IN ASTM FUELS A AND B

| | | | | | |
|---|---|---|---|---|---|
| PVC (K value of 70) | 100 | 100 | 100 | 100 | 100 |
| NBR (Chemigum P83) | 0 | 15 | 30 | 45 | 60 |
| DIDP | 77 | 69 | 67 | 65 | 63 |
| Ba/Cd stabiliser | 3 | 3 | 3 | 3 | 3 |
| Stearic acid | 1 | 1 | 1 | 1 | 1 |
| CaCO₃ | 10 | 10 | 10 | 10 | 10 |
| *Original physical properties* | | | | | |
| Hardness, Shore A | 72 | 74 | 73 | 71 | 70 |
| Modulus (MPa) | | | | | |
| At 100% | 8·0 | 7·6 | 6·8 | 6·1 | 5·5 |
| At 300% | 14·5 | 13·6 | 12·8 | 11·8 | 11·0 |
| Tensile strength (MPa) | 15·5 | 16·5 | 16·8 | 17·0 | 17·3 |
| Elongation (%) | 340 | 395 | 410 | 440 | 450 |
| Low-temp. flexibility (°C) | −40 | −37 | −35 | −34 | −33 |
| *Immersion in Fuel A: 70 h at 21°C* | | | | | |
| Hardness, Shore A | | | 91 | 72 | 69 |
| Low-temp. flexibility (°C) | 15 | 8 | −10 | −34 | −32 |
| Change in weight (%) | −26·6 | −23·7 | −12·3 | 0·5 | 0·5 |
| Change in volume (%) | −25·6 | −25·1 | −8·7 | 2·1 | 0·9 |
| Change in weight, dry (%) | −30·6 | −24·2 | −13·3 | −3·3 | −1·7 |
| Change in volume, dry (%) | −34·5 | −24·7 | −11·2 | −3·8 | −2·1 |
| *Immersion in Fuel B: 70 h at 21°C* | | | | | |
| Hardness, Shore A | | | | 92 | 88 |
| Low-temp. flexibility (°C) | 2 | −13 | −14 | −19 | −25 |
| Change in weight (%) | −11·3 | −7·2 | −2·8 | 4·0 | 11·7 |
| Change in volume (%) | −5·1 | −0·3 | 5·9 | 15·5 | 25·2 |
| Change in weight, dry (%) | −20·3 | −18·2 | −17·5 | −15·1 | −11·7 |
| Change in volume, dry (%) | −17·9 | −16·4 | −15·7 | −12·4 | −7·9 |

flexibility that occur after immersion in ASTM Fuel A (hexane) and ASTM Fuel B (70% iso-octane and 30% toluene) are shown in Table 9 and again demonstrate the effectiveness of NBR as a 'permanent' modifier.

### 5.7.3 Flexibility Resistance After Immersion
The usefulness of a PVC product can be drastically impaired due to loss of flexibility and cut growth resistance caused by contact with fuels, oils and fats. This is particularly true for PVC in industrial footwear applications.

Table 10 shows the increased resistance to cut growth after immersion in cutting oil, iso-octane and pig's fat.

TABLE 10

CUT GROWTH RESISTANCE AFTER IMMERSION
(ROSS FLEX AT −5°C AFTER 150 000 CYCLES)

|  | Test compounds | | |
|  | A | B | C |
|---|---|---|---|
| PVC (K value of 70) | 100 | 100 | 100 |
| NBR[a] | — | 20 | 30 |
| DOP | 88 | 82 | 80 |
| Epoxidised soya bean oil | 5 | 5 | 5 |
| Stabiliser/lubricant | 3·6 | 3·6 | 3·6 |
| Hardness, Shore A | 60 | 60 | 60 |
| Cut growth resistance (mm) | | | |
| Cutting oil, 70 days at 20°C | 9·6 | 1·95 | 1·20 |
| Iso-octane, 22 h at 21°C | (0·05)[b] | (35·0)[b] | 0·13 |
| Pig's fat, 7 days at 50°C | (1·53)[b] | (16·28)[b] | (22·01)[b] |

[a] NBR: ACN 33%, 80 $ML_4$, pre-crosslinked (Chemigum P8).
[b] Sample broke before test completed (kilocycles indicated in parentheses).

### 5.7.4 Oil and Fuel Resistance Comparison of NBR and Polymeric Plasticisers

Low-molecular-weight plasticisers like DOP are easily extracted by oils and solvents, and their use is limited; higher-molecular-weight polymeric plasticisers have been introduced to overcome this. The use of NBR with a monomeric plasticiser has been shown to be more efficient and cheaper on a volume basis than a monomeric/polymeric blend.

Table 11 lists the changes in physical properties that occur after immersion in ASTM Fuel B, gasoline and cooking oil.

### 5.8 Plasticiser Migration and Volatility[10,11,31]

The effectiveness of a plasticiser in its action in PVC is dependent on its ability to remain within the PVC compound; two physical characteristics that govern this are volatility and migration.

### 5.8.1 Volatility

The difference in volatility of NBR compared with a monomeric plasticiser both at equal loadings and compounded for equal hardness has been evaluated after seven days at 100°C in circulating air. Table 12 lists the results, showing a 10-fold reduction in volatile loss.

TABLE 11
### COMPARISON OF NBR AND POLYMERIC PLASTICISERS

|  |  |  |  |
|---|---|---|---|
| PVC | 100 | 100 | 100 |
| DOP | 75 | 50 | 75 |
| Polyester plasticiser | — | 25 | — |
| NBR (Chemigum P8) | — | — | 30 |
| Ba/Cd stabiliser | 3 | 3 | 3 |
| Stearic acid | 0·5 | 0·5 | 0·5 |
| Density | 1·19 | 1·22 | 1·16 |
| *Original physical properties* | | | |
| Hardness, Shore A | 61 | 61·5 | 57 |
| 300% modulus (MPa) | 10·7 | 11·7 | 8·9 |
| Tensile strength (MPa) | 14·2 | 15·7 | 14·8 |
| Elongation at break (%) | 430 | 445 | 505 |
| Compression set (%) | 69 | 69 | 60·5 |
| Resilience (%) | 27 | 18·6 | 26 |
| *Aged in Fuel B 72 h at room temperature and dried* | | | |
| Hardness change | +19 | +8·5 | 0 |
| Tensile strength retained (%) | 79 | 76 | 92 |
| Elongation at break (%) | 250 | 320 | 470 |
| Retained (%) | 58 | 72 | 93 |
| Weight change (%) | −30 | −0·2 | −5·4 |
| *Aged in gasoline 72 h at room temperature and dried* | | | |
| Hardness change | +25 | +17·5 | +10 |
| Tensile strength retained (%) | 140 | 95 | 117 |
| Elongation at break (%) | 310 | 300 | 455 |
| Retained (%) | 72 | 67 | 90 |
| Weight change (%) | −10 | −1·4 | −1·5 |
| *Aged in cooking oil 7 days at room temperature* | | | |
| Hardness change | +11 | +0·5 | 0 |
| Tensile strength retained (%) | 110 | 93 | 90 |
| Elongation at break (%) | 375 | 400 | 495 |
| Retained (%) | 87 | 90 | 98 |
| Weight change (%) | −5·3 | −1·6 | −1·8 |

### TABLE 12
### VOLATILE LOSS FROM NBR/PVC BLENDS

|  |  |  |  |  |
|---|---|---|---|---|
| PVC (K value of 70) | 100 | 100 | 100 | 100 |
| DOP | 100 | — | 40 | — |
| NBR | — | 100 | — | 80 |
| Ba/Cd stabiliser | 3 | 3 | 3 | 3 |
| Organic phosphite | 1 | 1 | 1 | 1 |
| Zinc stearate | 1 | 1 | 1 | 1 |
| Hardness, Shore A | 58 | 87 | 93 | 94 |
| Volatile loss (%) | 22 | 2 | 20 | 2 |

### 5.8.2 Migration

Plasticised PVC compounds when in contact with other plastic surfaces can lose plasticiser through migration. This is particularly true when plasticised PVC is in contact with a rigid PVC component. Here there is a tendency for the plasticiser to move from the plasticiser-rich environment to the plasticiser-'starved' compound. The inclusion of NBR in the plasticiser system reduces this migratory tendency, thereby maintaining the intrinsic properties of both components of a product.

Figure 9 shows the loss in milligrams (mg) from modified (A) and unmodified (B) PVC compounds placed in contact with a rigid PVC surface for up to 14 days at 70°C (DIN 53405).

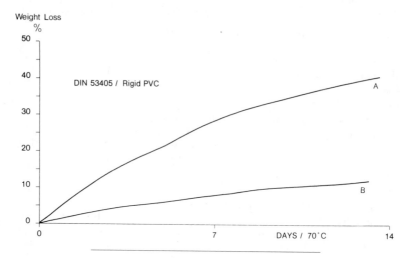

|  | Test compounds | |
|---|---|---|
|  | A | B |
| PVC | 100 | 100 |
| DOP | 65 | 65 |
| NBR[a] | 40 | — |
| $CaCO_3$ | 10 | 10 |
| Ba/Cd stabiliser | 3 | 3 |
| Stearic acid | 1 | 1 |

[a] NBR: 33% ACN, 50 $ML_4$ (Chemigum P83).

FIG. 9. Plasticiser migration study. A, Unmodified PVC; B, modified PVC/NBR compound.

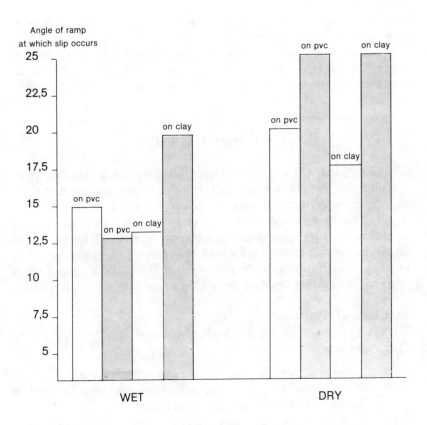

| | Test compounds | |
|---|---|---|
| | ☐ PVC | ☐ PVC/NBR |
| PVC | 100 | 100 |
| Plasticiser | 90 | 90 |
| NBR[a] | — | 50 |
| Ba/Cd stabiliser | 2 | 2 |
| Zinc stearate | 1 | 1 |

[a] Chemigum P612.

FIG. 10. Skid resistance study.

## 5.9 Skid Resistance

PVC compounds do not have good skid resistance and this has been a major defect in their application in the footwear industry. Skid tests on an inclined ramp have shown that the inclusion of NBR in a PVC compound increases the slip resistance in dry conditions on PVC and clay tiles, and in wet conditions on clay tiles. There is no improvement on wet PVC tiles (see Fig. 10).

# 6 COMPOUNDING

Compound design (choice of stabiliser, lubricant, filler, etc.) for NBR-modified PVC follows the same basic rules that apply for standard PVC, and the changes that can occur in physical properties when NBR is included have been discussed in Section 5. There are only two areas of compounding that require comment, namely (1) choice of NBR grade, and (2) the relationship between NBR loading and liquid plasticiser content in order to maintain the same flexibility as an unmodified compound, this property being closely linked to the apparent final hardness of the product.

## 6.1 Choice of NBR Grade

The available grades are listed in Table 2 in Section 3 and generally can be categorised according to:

(1) the molecular composition, i.e. the acrylonitrile content;
(2) the molecular weight, which determines the plasticity and is expressed in Mooney viscosity units ($ML_4$);
(3) the molecular configuration (either linear or pre-crosslinked).

### 6.1.1 Acrylonitrile Content

For maximum physical properties and oil and solvent resistance where compatibility is the key, polymers above 30% acrylonitrile content should be used.

However, if maximum improvements in unaged low-temperature flexibility and resilience are needed NBRs with acrylonitrile contents below 30% could be used, but compatibility will suffer, and mechanical properties will be reduced.

### 6.1.2 Mooney Viscosity ($ML_4$)

NBR polymers have high molecular weights and though they are thermoplastic they will increase the melt viscosity of PVC compounds at processing temperatures.

Therefore for processes that depend upon having low melt viscosities, such as injection moulding, polymers with Mooney viscosities of 40 and below should be used. High-acrylonitrile polymers (40% above) are more thermoplastic, but their impact on the melt processing is marginal. For the highest possible mechanical properties (tensile strength and modulus) high-viscosity polymers should be used. By contrast, for high elongation at break figures low-viscosity grades are preferred.

### 6.1.3 Molecular Configuration
Comparing linear polymers with pre-crosslinked types, the following trends apply.

Linear polymers give lower melt viscosities, lower change in hardness and higher elongations at break.

Pre-crosslinked polymers give the highest change in tensile strength, modulus and hardness, coupled with the highest reduction in compression set and abrasion loss. Pre-crosslinked polymers are also preferred for extrusion and calendering as their molecular structure imparts dimensional stability to the product, leading to more accurate dimensional control due to the reduced elastic memory of the NBR.

### 6.2 NBR/Plasticiser Relationship
If the monomeric plasticiser (e.g. DOP) content is kept constant there is a 1-point Shore A reduction in hardness for every additional 4–5 phr NBR added (depending on NBR types).

For substitution to give constant hardness the relationship is approximately three parts of NBR for one part of plasticiser if the polymer is pre-crosslinked and two parts NBR for one part plasticiser if it is linear.

## 7 PROCESSING

The means by which NBR-modified PVC compounds can be processed present no serious practical problems and mixing, extruding, injection moulding and calendering can all be successfully carried out on existing equipment. However, certain changes in operating conditions may be necessary and these are detailed in this section.

## 7.1 Mixing

The methods used for combining the PVC resin with the other compounding ingredients to obtain a homogeneous state suitable for future processing can be divided into two major classifications.

(1) Melt compounding, where the mix is completely fluxed and the compound is ready for final processing. This is normally carried out in an internal mixer of the 'Banbury' type or in continuous extruder-type mixers of various designs based on single- or multi-screw configurations.

(2) Dry blending, where the PVC and the liquid and dry compounding ingredients are combined together to form a free-flowing powder which is then subjected to heat and mechanical shear in a secondary process in order to obtain the completely fluxed compound ready for final processing.

Acrylonitrile–butadiene polymers can be mixed with PVC by both the above methods.

### 7.1.1 Melt Compounding

The melt compounding method involves higher capital expenditure. The fluxing of the PVC, plasticiser and NBR is carried out at temperatures of up to 170°C and the blend coming from the internal mixer has to be effectively and efficiently cooled, usually by 'dumping' on to a cooled two-roll mill, then strip-fed into a pelletiser from which the compound is stored ready for future use. The use of this type of mixing allows the NBR to be used in bale form—which is typical of rubber-type mixing.

### 7.1.2 Dry Blending

The real breakthrough in the use of NBR as a PVC modifier has come with the introduction of NBRs in powder form; this has allowed the polymer to be used in the less expensive, and more frequently used, powder dry blending equipment. The temperatures used are lower—up to 120°C—but to obtain good dispersion certain precautions have to be taken.

*7.1.2.1 Mixing technique.* For effective dry blending it is normal to use a two-stage mixing technique.

The first stage can be described as the 'dispersion stage' and here, with the help of heat obtained either by the speed of mixing or by jacketing, the PVC absorbs the liquid components and a free-flowing powder is obtained. Usually addition of the liquids is withheld until the PVC has attained a temperature of 70–80°C, at which point the PVC particles will absorb the liquids more easily. The temperature continues to rise during this first stage of mixing (up to 100–120°C) and to prevent reagglomeration and possible degradation the PVC blend is cooled in a low-speed second stage.

When using NBR powder as a modifier for a PVC compound, certain precautions must be taken. If the NBR powder is added to the PVC before plasticiser addition, the NBR will preferentially absorb the plasticiser, resulting in the formation of 'jelly-like' lumps that prevent the formation of a dry blend. For this reason it is essential that all the liquid plasticiser must be added to and absorbed by the PVC prior to the addition of the NBR.

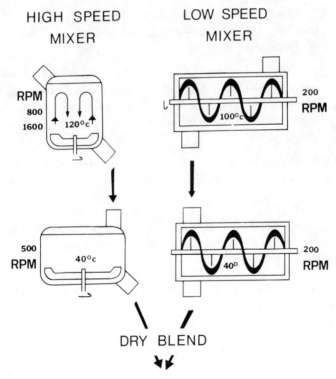

FIG. 11. PVC/NBR dry blending mixing procedures.

A typical dry blending procedure is as follows and is illustrated in Fig. 11.

(a)  High-speed mixing.

  (1)  Premix PVC and other dry ingredients except fillers at high speed until 70–80°C is reached.
  (2)  At this speed add liquid plasticisers.
  (3)  At 120°C add fillers and then dump into the cooling stage.
  (4)  Cool to 40°C and add the NBR powder.
  (5)  Discharge.

(b)  Low-speed mixing.

  (1)  Preheat the PVC and other dry ingredients except fillers to 70–80°C.
  (2)  Add the liquid plasticisers and let temperature rise to 100°C—add fillers.
  (3)  Dump into the cooling stage.
  (4)  Cool to 40°C and add the NBR powder.
  (5)  Discharge.

A useful practical test to determine if all the liquid plasticisers have been absorbed on to the PVC is to take a sample after step (2) and press between filter papers. If absorption is complete no trace of liquid plasticiser will be found on the filter papers—this is known as the 'dry point'.

*Note:* If for practical reasons it is not possible to add the NBR in the cooling stage it may be added in the first stage just prior to passing to the cooling stage.

*7.1.2.2 Gelation of the dry blend.* A correctly prepared PVC/NBR dry blend is completely free-flowing and will be stable in storage under normal conditions for several months.

Once the dry blend is prepared it can be fluxed (gelled), preferably in an extruder (though open, two-roll mills and internal mixers can be used), and then pelletised prior to final processing. This is the procedure normally followed by industrial compounders who supply compounded PVC/NBR blends to the final product manufacturer. It is also possible to pass directly from dry blend to finished product by feeding the dry blend directly to the injection moulding machine or extruder; by contrast, calendering always needs diced or pelletised stock.

The optimum temperature for fluxing the dry blend depends on the compound composition, and in particular the relative concentrations of conventional liquid plasticiser and NBR. The normal temperatures used vary from 140°C for the feed end of the extruder, progressing up to 170/180°C at the head. A check on dispersion can be carried out by milling a fine sheet at 160°C—any undispersed NBR will be immediately visible.

## 7.2 Extrusion

The extrusion of NBR-modified PVC blends poses no practical problems; the equipment normally used for PVC is suitable with $L/D$ ratios from 16:1 to 24:1 for single-screw and 7:1 to 15:1 for twin-screw being satisfactory. The temperature gradient required on the machine may need slight adjustment depending on the NBR/plasticiser ratio used, ranging from 130°C at the feed to 170/180°C at the die.

TABLE 13
MELT STABILITY CHARACTERISTICS

| | | |
|---|---|---|
| PVC ($K$ value of 70) | 100 | 100 |
| NBR (Chemigum P83) | 40 | — |
| DOP | 60 | 68 |
| TNP (trinonyl phosphate) | 0·5 | 0·5 |
| Ba/Cd stabiliser | 1·5 | 1·5 |
| Hardness, Shore A | 70 | 70 |
| *Profile extrusion test (140/150/160/170°C)* | | |
| Screw speed to melt fracture (rpm) | 100 | 40 |
| *Maclow–Smith Plastometer (ribbon die, piston speed 90 mm min$^{-1}$)* | | |
| Melt fracture temperature (°C) | 200 | 180 |

The inclusion of NBR in a PVC compound, however, does have some positive processing advantages in allowing higher temperatures and extruder speeds to be used before melt fracture occurs, as shown in Table 13.

## 7.3 Injection Moulding

NBR-modified PVC stocks can be easily injection moulded either directly from dry blends or from pelletised stock. A comparison of the conditions required to process has been extensively studied by De Marco et al.,[9] and the standard shoe sole stock results are reproduced in Table 14.

TABLE 14

INJECTION MOULDING MACHINE CONDITIONS FOR SHOE SOLE PRODUCTION FROM POWDER BLENDS

*Test compound* (*parts by weight*)·

| PVC | 100 |
|---|---|
| NBR | 40 |
| Plasticiser | 80 |
| $CaCO_3$ | 15 |
| Lubricant | 3 |
| Stabiliser | 2 |

| | *Machine name* | | |
|---|---|---|---|
| | New Britain Ankerwerk | Desma (12-station) | Menico (non-reciprocating 10-station) |
| Shot weight (g) | 114 | 300 | Two of 250 |
| Temperatures (start of run/end of run) (°C) | | | |
| Barrel, hopper zone | 165/163 | 171/171 | 160 |
| Barrel, stage 1 | 165/171 | 171/174 | 182 |
| Barrel, stage 2 | 171/177 | 171/174 | — |
| Nozzle | 177/182 | 171/177 | 182 |
| Mould, stationary side | 10/24 | 24 | 24 |
| Mould, ejector side | 10/24 | 24 | 24 |
| Stock | 171 | 177 | 182 |
| Cycle times (s) | | | |
| Screw forward travel time | 1 | 3 | 10 |
| Hold time, screw forward | 9 | 3 | — |
| Screw backward travel time | 5 | 7 | — |
| Screw back, hold time | 10 | — | — |
| Mould, cool time | 5 | 137 | 90 |
| Total cycle time | 30 | 150 | 100 |
| Pressure (psig) | | | |
| Injection pressure | 800 | 800 | — |
| | $(56.25\,kg\,cm^{-2})$ | $(56.25\,kg\,cm^{-2})$ | |
| Back pressure | 0 | 0 | — |
| Screw speed (rpm) | 145 | 138 | 145 |
| Screw compression ratio | 2.39 | 2.4 | — |
| Screw $L/D$ ratio | 12/1 | — | — |
| Screw diameter (in) | 2.125 | — | — |
| | (5.40 cm) | | |
| Nozzle outlet diameter (in) | $\frac{5}{16}$ | $\frac{5}{16}$ | — |
| | (0.80 cm) | (0.80 cm) | |

## 8 FINAL APPLICATIONS: STARTING FORMULATIONS

### 8.1 Footwear (Fig. 12)

|  | Unit sole, compact | Leisure sole, microcellular | Dairy boot | *Safety boot,[a] compact |
|---|---|---|---|---|
| PVC (K value of 65) | 100 | — | 100 | — |
| PVC (K value of 70) | — | 100 | — | 100 |
| NBR (Chemigum P83) | 20 | 20 | 50 | 30 |
| DOP | 80 | 80 | 80 | 80 |
| Calcium carbonate | 5 | 5 | — | — |
| Epoxidised soya bean oil | 3 | 3 | 12 | 5 |
| Stabiliser (barium/zinc) | 2·5 | — | 3 | 3 |
| Stabiliser (barium/lead) | — | 2·5 | — | — |
| Blowing agent (ADC) | — | 1·5 | — | — |
| Calcium stearate | 0·5 | 0·5 | 0·5 | 1·0 |
| *Physical properties* |  |  |  |  |
| Hardness, Shore A | 60 | — | 55 | 62 |
| Specific gravity | 1·16 | 0·75 | 1·12 | 1·16 |
| Abrasion (DIN) (mm$^3$) | 113 | 120 | 106 | 100 |

[a]Approved to DIN standard 4843 and BS 1870 Part 1.

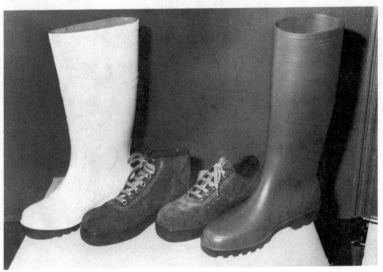

Fig. 12.   NBR-modified PVC footwear.

FIG. 13.  Hoses manufactured from NBR-modified PVC.

### 8.1.1 Compounds for Injection Moulding of Wellington Boots

|  | A (sole) | B (upper) |
|---|---|---|
| PVC (K value of 65) | 100 | — |
| PVC (K value of 60) | — | 100 |
| NBR (Chemigum P612) | 50 | 22·5 |
| Epoxidised soya bean oil | 10 | 20 |
| Polymeric plasticiser (Santiciser 409) | 12 | 35 |
| DOP (dioctyl phthalate) | 30 | 55 |
| Calcium carbonate (Socal D) | 10 | 5 |
| Stabiliser (Ba/Zn type) | 2·0 | 2·0 |
| Calcium stearate | 0·5 | 0·5 |

### 8.1.2 Injection Moulding Conditions for Desma 709 Ten-Station

|  | A (sole) | B (upper) |
|---|---|---|
| Zone 1 | 180°C | 185°C |
| Zone 2 | 190°C | 195°C |
| Zone 3 | 190°C | 195°C |
| Nozzle | 190°C | 180°C |
| Injection pressure | 22 kg cm$^{-2}$ | 44 kg cm$^{-2}$ |
| Plasticisation | 25 kg cm$^{-2}$ | 25 kg cm$^{-2}$ |
| Injection time | 10–12 s | 2·5 s |
| Screw speed | 110 rpm | 112 rpm |

## 8.2 Hoses (Fig. 13)

|  | General-purpose hose | High-performance hose | Tube resistant to oils and fuels |
|---|---|---|---|
| PVC (K value of 70) | 100 | 100 | — |
| PVC (K value of 75) | — | — | 100 |
| NBR (Chemigum P83) | 25 | 50 | 100 |
| DOP | 78 | 85 | 40 |
| Calcium carbonate | — | — | 20 |
| Epoxidised soya bean oil | 3 | 5 | 5 |
| Stabiliser (barium/lead) | 2 | 3 | 3·5 |
| Lubricant | 0·5 | 0·5 | 0·5 |
| *Physical properties* | | | |
| Specific gravity | 1·17 | 1·12 | 1·20 |
| Hardness, Shore A | 57 | 59 | 65 |

(continued)

|                                          | General-purpose hose | High-performance hose | Tube resistant to oils and fuels |
|------------------------------------------|:---:|:---:|:---:|
| Tensile strength (MPa)                   | 10·8 | 13·8 | 13·3 |
| Elongation at break (%)                  | 390  | 380  | 450  |
| Low-temp. flexibility (Clash & Berg) (°C) | −43  | −44  | −24  |
| Change in weight (%) after immersion in  |      |      |      |
| Iso-octane, 22 h at 20°C                 | +0·13 | —   | +2·1 |
| Oil ASTM 3, 24 h at 125°C                | —    | —    | −2·0 |
| Gasohol: 80% petrol/20% ethanol, 70 h at 20°C | — | — | +10·5 |
| Change in flexibility at low temp. (°C) after immersion in |  |  |  |
| Oil ASTM 3, 24 h at 125°C                | —    | —    | 0    |
| Gasohol, 70 h at 20°C                    | —    | —    | −3   |

## 8.3 Extruded Window Gaskets and Door Seals

|                                          | Solid | Sponge |
|------------------------------------------|:---:|:---:|
| PVC (K value of 70)                      | 100 | 100 |
| NBR (Chemigum P83)                       | 30  | 25  |
| DIDP                                     | 80  | 90  |
| Calcium carbonate                        | 10  | —   |
| Epoxidised soya bean oil                 | 3   | 4   |
| Stabiliser (barium/zinc)                 | 3·0 | 3·0 |
| Blowing agent (azodicarbonamide—ADC)     | —   | 1·75 |
| Calcium stearate                         | 0·5 | 0·5 |
| *Physical properties*                    |     |     |
| Specific gravity                         | 1·16 | 0·7 |
| Hardness, Shore A                        | 62  | —   |
| Compression set, 22 h at 70°C            | 59  | —   |
| Low-temp. flexibility (Clash & Berg) (°C) | −40 | —  |
| Low-temp. brittleness (°C)               | −60 | —   |

## 8.4 Cable Jacket

|                                          |     |
|------------------------------------------|:---:|
| PVC (electrical grade, K value of 70)    | 100 |
| NBR (Chemigum P83)                       | 40  |
| DOP (dioctyl phthalate)                  | 12  |

| | |
|---|---|
| DIOP (di-iso-octyl phthalate) | 30 |
| BBP (butyl benzyl phthalate) | 6 |
| DIDP (di-isodecyl phthalate) | 12 |
| Basic lead silicate sulphate | 4·0 |
| Calcined clay | 5·0 |
| Dibasic lead phosphite | 1·0 |

*Physical properties*

| | |
|---|---|
| Hardness, Shore A | 72 |
| Tensile strength (MPa) | 16·7 |
| Elongation at break (%) | 700 |
| Low-temp. brittleness (°C) | −37 |
| Volume resistivity ($\Omega\,cm^{-1}$) | $9\cdot0 \times 10^{11}$ |
| Volume resistivity after 10 days in water ($\Omega\,cm^{-1}$) | $1\cdot0 \times 10^{12}$ |

*Notes*
(1) The blend of monomeric plasticisers used gives excellent electrical properties.
(2) Basic lead silicate sulphate gives good heat, electrical and moisture repellency properties for PVC insulations.
(3) Calcined clay is used in place of calcium carbonate because of its superior electrical properties.
(4) Dibasic lead phosphite is used as a light and weathering stabiliser for low-temperature PVC electrical insulations.

## 8.5 Crash-pad Skin (Fig. 14)

*Typical crash-pad formulation*

| | |
|---|---|
| PVC (*K* value of 65) | 100 |
| ABS (Blendex 101) | 70 |
| NBR (Chemigum P8) | 30 |
| DOP | 20 |
| DIDP | 20 |
| Epoxidised soya bean oil | 10 |
| Stabiliser (barium/zinc) | 2·0 |
| Stearic acid | 0·5 |
| Carbon black | 1·0 |

## 8.6 Sealing Strip for Refrigerator Doors

| | |
|---|---|
| PVC (suspension type, *K* value of 70) | 100 |
| NBR | 25 |
| Stabiliser (barium/zinc) | 2·0 |

Fig. 14.    Vacuum-formed PVC/ABS/NBR crash pad.

| | |
|---|---|
| Stearic acid | 0·2 |
| Calcium carbonate (Omya BSH) | 15 |
| Titanium dioxide (rutile) | 2 |
| BBP (butyl benzyl phthalate) | 35 |
| DOA (dioctyl adipate) | 15 |
| Polymeric plasticiser (Santiciser 409A) | 20 |

*Physical properties*

| | |
|---|---|
| Hardness, Shore A | 66 |
| Tensile strength (MPa) | 16·3 |
| Elongation at break (%) | 840 |
| Compression set (%) | 25 |
| Tension set (%) | 24 |
| Low-temp. brittleness (°C) | −41·5 |
| Low-temp. brittleness after ageing 7 days at 100°C (°C) | −36 |

## 9 CONCLUSIONS

NBR-modified PVC compounds have been used on an industrial scale to produce the following products, the majority of which were previously manufactured from vulcanised rubber.

*Extrusion Process*

| Profiles | Window gaskets and lip seals |
| | Door seals |
| | Refrigerator gaskets |
| | Cellular seals |
| | Water stops and dampcourse strips |
| | Co-extruded automotive gaskets |
| | Co-extruded building profiles |
| Tubes | Automotive hoses and tubing |
| | Gas tubing |
| | Chemical tubing |
| | Co-extruded spiral hoses |
| Cable jackets | Non-migratory sheaths |
| | Welding cable covers |
| | Industrial cable jackets |

*Injection Process*

| Footwear | Safety boots and shoes |
| | Leisure soling |
| | Sports soling |
| | Golf shoes |
| | Microcellular soling |
| | Dairy boots |
| | Industrial boots |
| | Non-slip soling |
| Automotive | Protective covers |
| | Grommets |
| | Electrical plugs |
| | Fixed-glass gaskets |

*Calendering and Thermoforming*

| Automotive | Insulating sheets |
| | Embossed panels |
| | Crash pads |
| Industrial | Phonic insulation |
| | Flooring |
| | Food packaging |

*Extrusion Blow Moulding*

| Convoluted covers | Automotive shaft covers |
| | Industrial bellows |
| | Cable sleeves |

## ACKNOWLEDGEMENT

The author wishes to thank the Goodyear Tire and Rubber Co. for permission to publish this paper.

## REFERENCES

1. SEMON, W. L. (Goodrich BF), US Patent 1929 453, 10 October 1933.
2. SEMON, W. L. (Goodrich BF), US Patent 2188 396, 30 January 1940.
3. BAUMANN, E., *Liebigs Ann.*, **163** (1872) 308.
4. LAWSON, W. E. (DuPont de Nemours), US Patent 1 867 014, 12 July 1932.
5. REID, E. W. (Union Carbide), US Patent 1 935 577, 14 November 1933.
6. VOSS, A. & DICKHAUSER, E. (IG Farben), US Patent 2 012 177, 20 August 1935.
7. PEDLEY, K. A., *Polymer Age*, **1** (May 1970) 97.
8. GIUDICI, P. & MILNER, P. W., Nitrile rubber powders in PVC compounding, Paper presented at the *Technology of Plastics and Rubber Interface Conference*, PRI, Brussels, 1976.
9. DE MARCO, R. D., WOODS, M. E. & ARNOLD, L. F., *Rubber Chem. Technol.*, **45**(4) (1972) 1111.
10. ANON., *Hycar Nitrile Rubbers at PVC Modifiers*, Information Bulletin E141-8, Chemische Industrie AKU–Goodrich BV, Arnhem, The Netherlands.
11. ANON., *Chemigum Powder*, Technical Literature 1–8 and Technical Bulletins, Goodyear Chemicals Europe, Avenue des Tropiques, ZA de Courtaboeuf, 91952 Les Ulis Cedex, France.
12. SCHWARZ, H. F. & BLEY, J. W. F., Paper No. 33, ACS Rubber Div. 128th Meeting, Cleveland, Ohio, 1–4 October 1985.
13. PICKETT, W., *Rubber Age*, **103**(6) (1971) 69.
14. OKUNO, M., *Nippon Gomu Kyokaishi*, **48**(5) (1975) 314.
15. BYRNE, P. S. & SCHWARZ, H. F., *Rubber Age*, **105**(7) (1973) 43.
16. SIMMONS, P., *Rubber & Plastics Age*, **48**(5) (1967) 442.
17. ZAKRZEWSKI, G. A., *Polymer*, **14** (1973) 348.
18. NISHI, T. & KWEI, T. K., *J. Appl. Polym. Sci.*, **20** (1976) 1331.
19. HUBELL, D. S. & COOPER, S. L., *J. Polym. Sci., Polym. Phys. Ed.*, **15** (1977) 1143.
20. THOMAS, D. A., *J. Polym. Sci., Polym. Symp.*, **60** (1977) 189.
21. AGGARWAL, S. L., *Polymer*, **17** (1976) 938.
22. NISHI, T., KWEI, T. K. & WANG, T. T., *J. Appl. Phys.*, **46** (1975) 4157.
23. NISHI, T. & KWEI, T. K., *J. Appl. Polym. Sci.*, **20** (1976) 1331.
24. SAUER, J. A. & WOODWARD, A. E., *Rev. Mod. Phys.*, **32** (1960) 88.
25. COLEMAN, M. M. & VARNELL, D. F., *J. Polym. Sci., Polym. Phys. Ed.*, **18** (1980) 1403.
26. KLEINER, L. W., KARASZ, F. E. & MACKNIGHT, W. J., *Soc. Plast. Eng. Tech. Pap.* (1978) 243.
27. STOELTING, J., KARASZ, F. E. & MACKNIGHT, W. J., *Polym. Eng. Sci.*, **10** (1970) 133.
28. WANG, C. B. & COOPER, S. L., *J. Polym. Sci., Polym. Phys. Ed.*, **21** (1983) 11.

29. HORVATH, J. W., WILSON, W. A., LUNDSTROM, H. S. & PURDON, J. R., *Appl. Polym. Symp.*, **7** (1968) 95.
30. BERGMAN, G., BERTUILSSON, H. & SHUR, Y. J., *J. Polym. Sci., Polym. Phys. Ed.*, **21** (1977) 2953.
31. DUVAL, G. R. & TANDON, P. K. (Goodyear Chemicals Europe), Les propriétés et les applications des mélanges PVC caoutchouc nitrile, paper presented at GPCP Conference, Paris, 1986.

*Chapter 4*

# CONTROL OF POLYMER PROCESSES

Keith T. O'Brien*

*Celanese Engineering Resins, Chatham,
New Jersey, USA*

## 1 INTRODUCTION

In the field of polymer processing, polymer feedstocks are converted into useful products. The methods of conversion are manifold and include extrusion, injection molding, blow molding, thermoforming, rotational molding and compression molding. The process parameters which influence these processes include temperature, speed, pressure, thickness and position. To make a successful product it is necessary to control these temperatures, speeds, pressures, thicknesses and positions. If they are not controlled, many parts will be rejected as substandard, and the performance of many others will be less than might have been. The degree of control depends upon how demanding the end-use specifications are. For example, compounded pellets formed to tight tolerances are unnecessary. However, blown film must be of uniform thickness over a small gage and tight dimensional tolerances must be held.

Each situation will demand a determination of the level of sophistication required, based upon business rather than technical data. With this in mind the various control strategies will each be treated here with emphasis on temperature, thickness, pressure, speed, position and time, in that order. However, before these treatments are made, basic control actions and their verifications will be explained, with particular reference to temperature control.

* Present address: New Jersey Institute of Technology, University Heights, Newark, New Jersey 07102, USA.

## 2 BASIC TEMPERATURE CONTROL ACTIONS

There are many control actions, and those most used in plastics processing will be treated here. The simplest will be dealt with first, followed by the more sophisticated.

### 2.1 On–Off Control

In on–off control the manipulated variable is only adjusted when the setpoint is crossed. In temperature control this means that if the actual temperature is below the setpoint value then the heater is on, and if it is above the setpoint value then the heater is turned off. When the heater is on, it is fully on, and when it is off, it is fully off. As the actual temperature crosses the setpoint, either from above or below, the heater comes on or switches off respectively. Thus the actual temperature is continually cycling, as shown in Fig. 1. The range of actual temperature, measured from peak to valley, and the period of the cycle, measured as the time between successive peaks, depend upon the process characteristics, primarily the rate of heating and cooling of the heater. The transfer curve for this type of control action is presented in Fig. 2, assuming that once the heater is on it is instantaneously at fully on and that once it is off it is instantaneously at fully off. This cannot happen in reality. Also, since for this ideal situation the response time would be zero, rapid cycling would occur. So it is clear

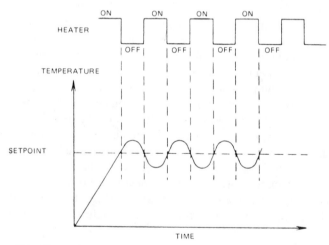

FIG. 1.   On–off temperature control action. (Courtesy of West Division, Gulton Industries Inc.)

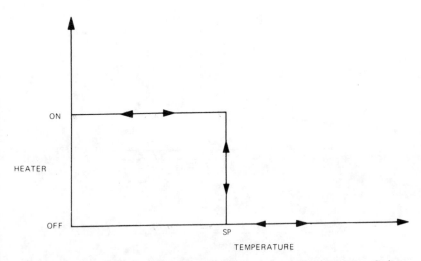

FIG. 2.   Ideal transfer curve for on–off control. (Courtesy of West Division, Gulton Industries Inc.)

that the temperature cannot be held at setpoint but can only be held near it. How near depends upon the control systems used. The length of the cycle also depends upon the control system but it is irrelevant as far as holding setpoint is concerned. In fact, fast cycles are detrimental to the control hardware and are usually avoided. To prevent fast cycles an on–off differential, or hysteresis, is built into the controller. This hysteresis requires that the actual temperature not only passes the setpoint but actually exceeds it positively or negatively by a certain amount before the control action is activated. The hysteresis will also damp out signal noise and prevent the output from chattering. Indeed the amount of hysteresis should be determined by the level of signal noise. It should be great enough to prevent chattering but small enough that the controlled temperature range is maintained within acceptable limits. Then the control action will be as shown in Fig. 3, which shows that the heater is switched on and off as the actual temperature crosses the hysteresis limits. Now the transfer curve changes to that shown in Fig. 4, where the actual temperature is initially at A with the heater on. The temperature rises until it reaches the upper hysteresis limit at B, when it turns off at C. The actual temperature continues to rise because the heater is still hot although cooling, D, and then drops until the lower hysteresis limit is reached at E. The heater then turns on, F, but the temperature continues to fall to point G since the heater is still

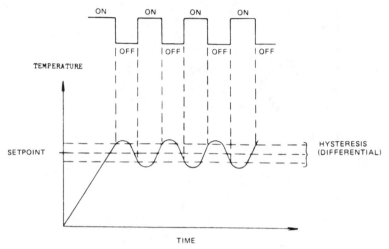

FIG. 3. On–off controller with hysteresis. (Courtesy of West Division, Gulton Industries Inc.)

warming up. The temperature rises until it reaches the upper hysteresis limit, and the cycle continues to repeat.

## 2.2 Proportional Control

An improvement to on–off control is proportional control. Taking the same temperature control example as before, in proportional control the amount of heat supplied by the heater to the extruder barrel is proportional to the temperature difference between the actual temperature and the

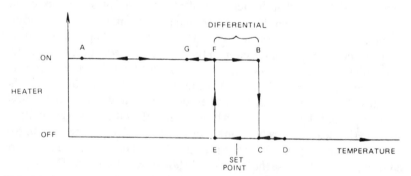

FIG. 4. On–off control: practical transfer curve with hysteresis. (Courtesy of West Division, Gulton Industries Inc.)

setpoint temperature. Using electric heaters, the proportional controller adjusts the power to the heater so that the heater supplies the barrel with the exact amount of heat required to maintain a stable temperature. The range over which the power is adjusted from fully off (0%) to fully on (100%) is termed the proportional band. If the actual temperature falls outside the proportional band then the heater is either fully on or fully off. The proportional band is typically expressed as a percentage of the instrument span, and is typically centered around the setpoint, with which it changes automatically. Thus, for a typical plastics processing temperature controller, the span would be 600°F (333°C) and a 5% proportional band would be 30°F (15°C) wide, 15°F (7·5°C) above and 15°F (7·5°C) below setpoint. A proportional controller is reverse-acting since the output decreases with increasing temperature. The negative slope of the output curve shown in Fig. 5 indicates this point. In this figure the power output of

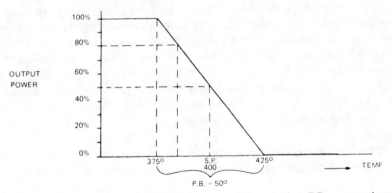

FIG. 5.   Proportional controller transfer curve. S.P., setpoint; P.B., proportional band. Temperature in °F. (Courtesy of West Division, Gulton Industries Inc.)

the heater is plotted against temperature for a situation where the setpoint is 400°F (205°C) and the proportional band is 50°F (26°C) (375–425°F, 192–218°C). Below a temperature of 375°F (192°C) the heater power is fully on (100%). Above a temperature of 425°F (218°C) the heater power is fully off (0%). Between these limits there is a linear relationship (proportional) between the power output and the temperature. The width of the proportional band is important because it influences the relationship between the power output and the temperature. The band may be adjusted to suit the needs of the various polymer processes. If the controller is tuned to have a wider proportional band, as shown in Fig. 6, then a large

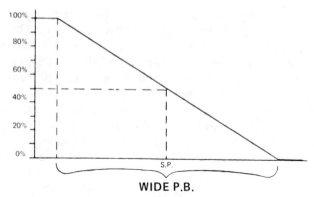

**WIDE P.B.**

FIG. 6.    Effect of wide proportional band (P.B.); S.P., setpoint. (Courtesy of West
Division, Gulton Industries Inc.)

temperature change is required to produce the same change in heater
power. The converse is also true. If the controller is tuned to have a narrow
proportional band, as shown in Fig. 7, then only a small temperature
change is required for the same change in heater power. In the limit, if the
proportional band were reduced to zero, then an on–off controller would
result. The gain of a proportional controller can be calculated by dividing
100 by the percentage of the proportional band. Fast responses, from
narrow bandwidths, result from high gains, which also lead to noise-
induced instabilities.

The block diagram of a proportional controller is presented in Fig. 8. The
temperature signal from the thermocouple is amplified, and used to drive a

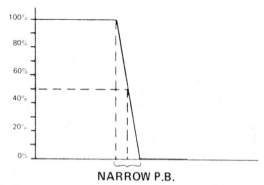

**NARROW P.B.**

FIG. 7.    Effect of narrow proportional band. (Courtesy of West Division, Gulton
Industries Inc.)

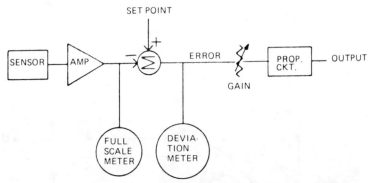

FIG. 8.   Proportional controller. PROP. CKT, Proportioning circuit. (Courtesy of West Division, Gulton Industries Inc.)

full-scale meter, giving an operator a chance to view the actual temperature. Cold junction compensation circuitry is included in the amplifier. The setpoint is determined by the operator, usually on a dial. The setpoint temperature and the actual temperature are compared in a comparator, and the difference, whether positive or negative, is the error or deviation signal. Ideally it is zero, and the heater will continue to supply exactly the right amount of heat. In practice, if it is positive the extruder barrel is below setpoint, and if it is negative the extruder barrel is above setpoint.

The error signal is input to the proportioning circuit after gain modification, and the output from the proportioning circuit is used to control the power to the heater. If the actual temperature coincided with the setpoint temperature then the heater power would be 50% of maximum. If the process was stable under these conditions they would remain. However, it would be unusual for the heat required to be 50% of maximum and in practice the temperature would increase or decrease from setpoint. This would cause the power to vary until equilibrium was reached, at the stabilized temperature. The difference between this temperature and the setpoint temperature is called the offset. If the process is under control then the stabilized temperature will lie within the proportional band. So the offset can be reduced by narrowing the proportional band to some point before instabilities arise. This is clearly seen in Fig. 9, which shows a temperature–time curve for a controller in a plastics process coming up to temperature. To understand better their inherent offset, imagine the temperature controller transfer curve (Fig. 5) superimposed on the process transfer curve, as shown in Fig. 10. In this particular example the zone of the plastics process is being heated with a 2-kW heater which produces a

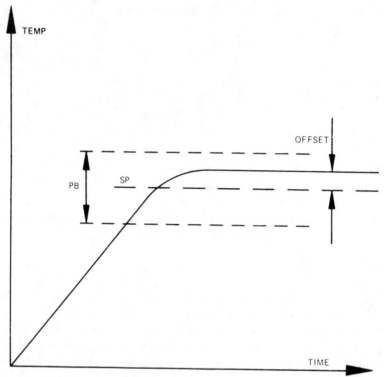

FIG. 9. Process with temperature offset. (Courtesy of West Division, Gulton Industries Inc.)

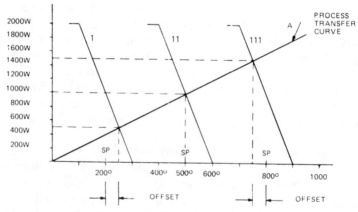

FIG. 10. Process offset illustration. Temperature in °F. (Courtesy of West Division, Gulton Industries Inc.)

zone temperature of 1000°F (538°C) when fully on. For illustrative purposes the relationship between heat input and zone temperature, the process transfer curve, has been assumed linear. Also included in this graph are three transfer curves for a 200°F (93°C) proportional band controller. These curves are for three different setpoints, 200°F (93°C), 500°F (260°C) and 800°F (427°C), and are marked 1, 11 and 111 respectively. Curve 1, for the 200°F (111°C) setpoint, intersects the process transfer curve at a power level of 500 W and a process temperature of 250°F (121°C). The offset is 50°F (27°C) positive. However, curve 11, for the 500°F (260°C) setpoint, intersects the process transfer curve at a power level of 1000 W and a process temperature of 500°F (260°C), with no offset. Finally, curve 111, for the 800°F (427°C) setpoint, intersects the process transfer curve at a power level of 1500 W and a process temperature of 750°F (399°C). The offset is 50°F (27°C) negative. So the offset can be seen to be dependent upon the process transfer curve, the proportional band (or gain) and also the setpoint. It is clear that proportional controllers are unsuitable for machines such as extruders and injection molding machines where setpoint movement is commonplace.

## 2.3 Manual and Automatic Reset Control
The offset inherent to proportional controllers can be removed manually or automatically. To reset the offset manually, the potentiometer is adjusted to offset electrically the proportioning band. The adjustment must be made in increments over a period of time to allow the process to stabilize under the new control conditions. It is a tedious procedure, but eventually the offset will reduce to zero. This process is illustrated graphically in Fig. 11, where

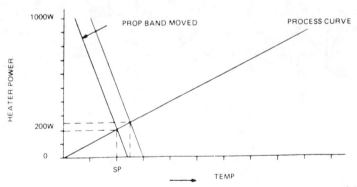

FIG. 11.    Manual reset action: heater power requirement. (Courtesy of West Division, Gulton Industries Inc.)

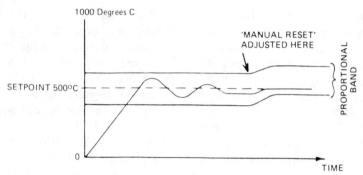

FIG. 12.   Manual reset action: temperature characteristic. (Courtesy of West Division, Gulton Industries Inc.)

the controller and process transfer curves are superimposed. To reach the setpoint the power output from the heater had to be reduced by 200 W. In Fig. 12 the same adjustments are shown on a time–temperature graph, where the proportional band has shifted to correct the offset. The block diagram for this control action becomes more complex and is shown in Fig. 13, where a second circuit (comparator) has been added prior to the proportioning circuit.

   Naturally, manual reset is both uncommon and unnecessary in the modern world of cheap electronics and automatic reset is usual in modern controllers. An electronic integrator is used to perform the reset automatically. The deviation signal is integrated with respect to time and

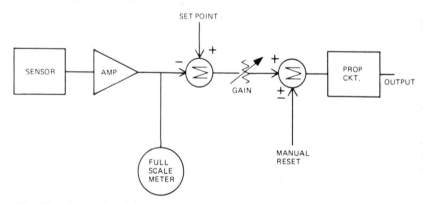

FIG. 13.   Proportional controller with manual reset. (Courtesy of West Division, Gulton Industries Inc.)

then summed with the deviation signal. This summation moves the proportional band owing to the inflated or deflated output values that result. The heater power automatically increases or decreases to bring the process back to the true setpoint. The integrator continually manipulates the heater power, and therefore the temperature, until the deviation is zero. At this condition the input to the integrator is zero, and its output stabilized with the correct amount of reset to maintain the process at setpoint. If the temperature requirements of the process change, another deviation will occur, and the integrator will again determine the reset necessary to reduce the offset to zero and then maintain it at zero. In fact, the integral term in the control action continually acts to reduce the offset to zero. It is very important, however, to apply the corrective action slowly so that the response speed of the controller to load is not exceeded. If it is not, then oscillations will manifest. The time constant of the integral action determines the speed of response. It is defined as the time interval in which that part of the output signal due to integral action increases by an amount equal to that part of the output signal due to proportional action, when deviation is constant. If a step change of 10% is made in the setpoint, the output will immediately increase, as shown in Fig. 14, due to the proportional action. However, the same increase due to the integral action is not reflected for another 5 min. Thus the reset time constant is 5 min. The block diagram for this proportional-plus-integral controller is presented in Fig. 15, showing the addition of the integrator, which integrates the deviation signal. The summing circuit then adds the deviation signal to this integral signal. Note that the gain is external to these actions, and also that automatic reset is frequently expressed in repeats per minute rather than the inversely proportional integral time constant.

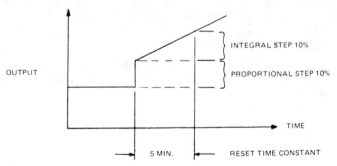

FIG. 14.  Reset time definition. (Courtesy of West Division, Gulton Industries Inc.)

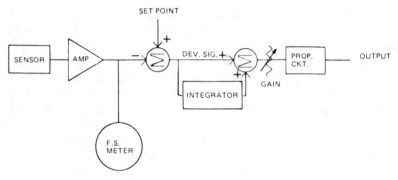

FIG. 15. Proportional-plus-integral controller. (Courtesy of West Division, Gulton Industries Inc.)

So much for the advantages of automatic reset. Unfortunately there are also some drawbacks. The most serious is integral saturation, which is the term used when an integrator has acted on a deviation signal when the actual temperature is outside the proportional band. This causes the proportional band to shift so far that the setpoint is now outside the band. Now the actual temperature must pass the setpoint before the controller output can change. When this eventually happens, the polarity of the deviation signal changes, and the output from the integrator begins to decrease (desaturate). The net effect of these occurrences is a very large temperature overshoot, which can be avoided by preventing the integrator from acting if the temperature is outside the proportional band. This is known as integral lockout, and is fairly standard on proportional-plus-integral controllers. Another characteristic of all such controllers is overshoot during start-up. This occurs because the integral lockout prevents the integrator from acting until the actual temperature has reached the bottom of the proportional band. By the time the temperature reaches the setpoint the integral action has already adjusted the proportional band upwards. The heaters respond by providing additional heat, giving rise to a temperature overshoot. This will correct itself over time because of the reset action of the integrator. However, lineout takes longer. A full graphical explanation of these actions is presented in Fig. 16, on a time–temperature plane.

### 2.4 Rate (or Derivative) Control
Rate control may be added to proportional control, as shown in the block diagram of Fig. 17. The rate controller determines the deviation of the

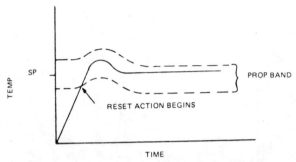

FIG. 16. Temperature characteristic for a proportional-plus-integral controller. (Courtesy of West Division, Gulton Industries Inc.)

signal, also known as the rate of change of the signal, with respect to time. This entity is then added to the proportional signal in a summing circuit and used to control the process. The rate action provides the controller with the ability to shift the proportional band to compensate for rapidly changing temperatures. Expressed in an analogous way to proportional-plus-integral control, it is proportional to the derivative time constant, which can be defined with reference to Fig. 18. It is the time interval required for that part of the output signal due to proportional action to increase by an amount equal to that part of the output signal due to derivative action, when the deviation is changing at a constant rate. This definition is made clearer by reference to Fig. 18. The time constant should be large enough to avoid the controller reacting to noise, which would cause a rapidly oscillating output signal. The rate action causes the gain of the controller to increase during temperature changes. This compensates for the lag in the process, and allows a narrower proportional band with less

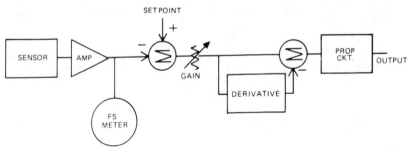

FIG. 17. Proportional-plus-rate controller. FS METER, Full scale meter; PROP. CKT, proportioning circuit. (Courtesy of West Division, Gulton Industries Inc.)

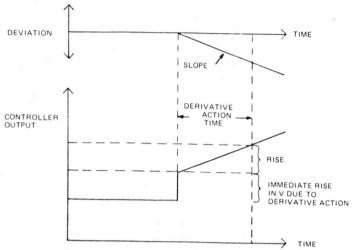

FIG. 18.    Derivative time definition. (Courtesy of West Division, Gulton Industries Inc.)

offset to be selected. The rate control action is not limited to applicability within the proportional band, and will operate without problems over the entire controller range. Since it operates off the temperature–time gradient it will help reduce overshoot upon start-up.

## 2.5 Proportional, Integral and Derivative (PID) Control

The advantages of proportional-plus-integral and proportional-plus-derivative control can be synthesized into PID control, which is

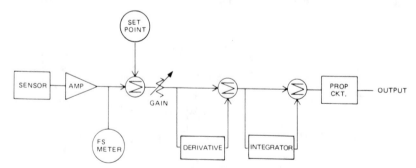

FIG. 19.    Proportional + integral + derivative controller. (Courtesy of West Division, Gulton Industries Inc.)

commonplace in the most difficult-to-control processes. The block diagram for a PID controller, which is also known as a three-mode controller, is presented in Fig. 19. This type of controller, when correctly tuned, will ensure that the temperature approaches the setpoint speedily and smoothly, without overshoot. It must be tuned so that the proportional-plus-rate signal from which the integrator works will be just sufficient for it to store the required integral value as the temperature reaches setpoint.

## 3 TYPES OF TEMPERATURE CONTROL STRATEGIES

Now that the basics of temperature control have been explained, the more complex practical systems can be discussed.

### 3.1 Temperature Control with Heating and Cooling

Many plastics processes, particularly extrusion, can be exothermic due to the mechanical working of the highly viscous materials. This generates heat, through viscous dissipation, which may require removal to maintain the setpoint temperature. So both heating and cooling control is required. A PID controller for such a situation is shown in Fig. 20, and the transfer function is shown in Fig. 21. Below the proportional band, 100% heating occurs. Above the proportional band, 100% cooling is applied. Within the proportional band, shown in Fig. 21, there is a linear reduction of heating to zero, and a linear increase of cooling from zero. But these actions are offset and only occur simultaneously over a small range of overlap. The overlap creates a smooth transition from heating to cooling and vice versa, which reduces overshoots. Another feature is the variation of cooling gain $X_{p2}$

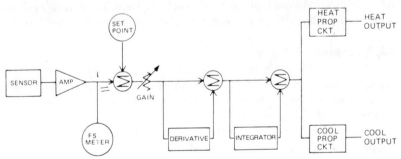

FIG. 20.   Heat–cool PID control. (Courtesy of West Division, Gulton Industries Inc.)

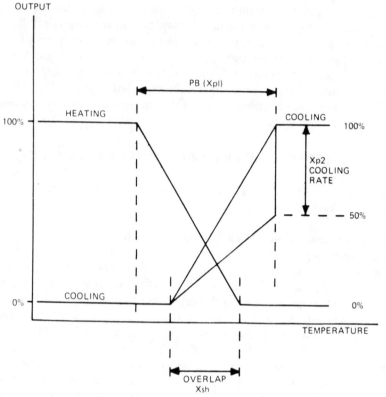

FIG. 21. Heat–cool temperature characteristic. (Courtesy of West Division, Gulton Industries Inc.)

which allows additional tuning options in the system. Temperature control with heating and cooling is utilized in intensive mixers and reactors.

### 3.2 Anticipatory Temperature Control

Inadequate temperature control causes product quality variations, and may precipitate instabilities. The importance of these effects depends upon the product, e.g. blown film requires close control whereas compounded pellets do not. Where close control is required, then small sources of change, which include minor screw speed variations, inconsistent polymer supply, regrind size variations, feedstock temperature variations and pressure variation due to clogging screenpacks, all become significant.

Specifically for extrusion, total temperature control packages have been introduced; they include anticipatory control, which as the name suggests reacts in the present to future changes of a variable. The anticipatory control action is possible by using two thermocouples, one close to the barrel wall and the other close to the heater. This is shown in Fig. 22. This type of temperature controller is also known as a deep well–shallow well type. The two thermocouple signals are input to a controller whose logic is based on an algorithm that accounts for the linear variation of temperature between the two thermocouples. In this way the controller reacts not only to the temperature on the barrel surface but also to the heat already conducting to that surface and its anticipated arrival time.

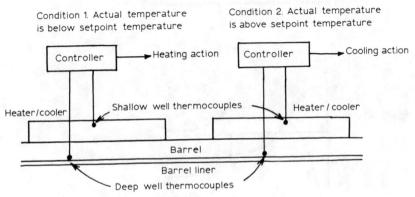

FIG. 22.   Anticipating temperature control.

For the circuit which is heating in Fig. 22, the transfer curve is shown in Fig. 23(a). The heater output diminishes and/or arrests in advance of the barrel wall reaching setpoint (2), because the controller has already anticipated that enough heat is conducting to that surface, based on the two input algorithms from which it works. The controller also has an automatic reset to eliminate offset (4), which activates as the deep-well temperature stabilizes. This shifts the setpoint and brings the actual temperature close to the actual setpoint (5).

For the circuit which is cooling in Fig. 22, the transfer curve is shown in Fig. 23(b). The heater output changes (2) to a cooler output in anticipation of the overshoot. Then automatic reset is introduced to eliminate the offset (4), and the actual temperature decreases to the actual setpoint temperature.

Systems such as the one just described, which is custom-built for

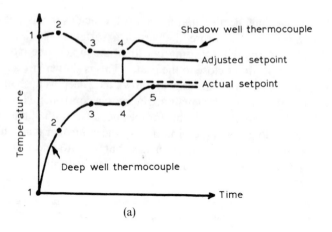

(a)

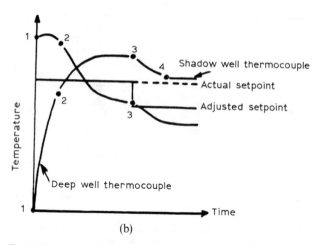

(b)

FIG. 23.   Transfer curves for anticipatory control: (a) heating load condition; (b) cooling load condition.

extrusion temperature control, offer faster cycling of heating and cooling actions, do not need any manual adjustment once tuned, have high gain in both heating and cooling circuits to ensure high thermal transfer rates, and can control more powerful heaters and coolers. In specific situations screw speeds may be increased with extra cooling and temperature stability maintained. The result is more output at equivalent quality. Material savings and energy savings are also possible.

### 3.3 Closed-loop Melt Temperature Control

Although close control is desirable on each zone so that the actual temperature is the setpoint temperature, the really important temperature to control is the melt temperature at entry to and through the die. If the extruder supplies melt with a variable stock (melt) temperature, then the product will vary in dimensions and physical properties. The real question is how much variation can be tolerated before an unacceptable product is produced. In compounding wide variations are acceptable because it is not critical that pellet diameters should be the same. With blown film, however, gage is of paramount importance so the right stock (melt) temperature control is required. To achieve closed-loop melt temperature control, a melt thermocouple must be installed between the extruder and the die. The signal from this thermocouple may then be used to adjust the setpoints of the zone controllers. Usually the setpoint of the zone controller immediately prior to the adaptor and die would be chosen, or if the adaptor is large then its controller setpoint could be changed. However, these setpoint changes must be made with caution since the extruder zones have been designed for specific purposes, and these should not be compromised. This process is illustrated in Fig. 24.

The melt temperature is measured by the melt thermocouple, and the signal input to the closed-loop controller. The control action of this controller is then used to change the setpoint of the die, the adaptor and the last barrel zone. The changes in setpoint will then precipitate changes in the melt temperature through the die.

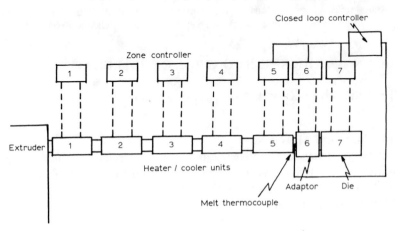

FIG. 24. Closed-loop temperature control.

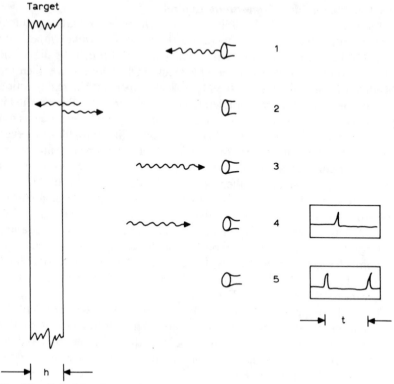

FIG. 25. Principle of thickness measurement using ultrasound. 1, Signal emitted; 2, signal reflects from first surface and transmits; 3, transmitted signal reflects from second surface; 4, first reflection received; 5, second reflection received.

## 4 THICKNESS CONTROL OF EXTRUDATES

There are two modes of thickness measurement used for thickness control in extrusion.

### 4.1 Ultrasound Measurement

Ultrasonic signals are emitted and aimed at the extrudate whose thickness must be measured. Some ultrasound is reflected from the surface of the material, whereas some is transmitted through the part and then reflected from the inner surface. These two reflected signals are sensed by an ultrasound receiver as they return. The first returns prior to the second, which has travelled further. The time interval between the reflected signals

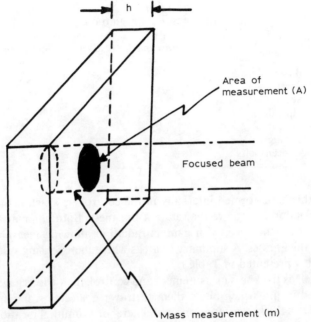

FIG. 26.    Principle of thickness measurement using nuclear gages.

can be measured as shown in Fig. 25. The distance travelled by the second signal during this time, $t$, is twice the extrudate thickness, $2h$. With prior knowledge of the velocity of sound in the plastic under measurement conditions, $V^s$, the thickness may be calculated from

$$h = V^s t/2 \qquad (1)$$

Of course this calculation is carried out in the controller, displayed for the operator, and control action is taken to change $h$ to the setpoint value.

## 4.2 Nuclear Radiation Measurement

The second thickness measurement technique is the use of nuclear radiation, where the mass of material, $m$, is measured in a region to which the sensor has been tuned, as in Fig. 26. The area for which the sensor has been tuned, $A$, is known, so if the density, $\rho$, of the extrudate under measurement conditions is known, then the thickness, $h$, may be calculated from

$$h = m/\rho A \qquad (2)$$

### TABLE 1
THICKNESS GAGING METHODS

|  | Ultrasound | Beta ray | Gamma ray |
|---|:---:|:---:|:---:|
| Blown film | √ | √ | √ |
| Sheet, thick | √ |  | √ |
| Sheet, thin | √ | √ | √ |
| Pipe, thick | √ |  | √ |
| Pipe, thin | √ |  | √ |
| Tubing | √ | √ | √ |
| Concentricity | √ | √ | √ |

Again this is performed internally in the controller, which facilitates the control action to achieve the setpoint thickness. Both gamma- and beta-radiation can be used, with gamma-radiation allowing measurement of greater thicknesses. A summary of areas where these sensing systems may be used is presented in Table 1.

As well as thickness measurement and control, the various gages may be mounted in jigs which allow them to traverse sheets or circumnavigate pipes. Thus thickness variations with lateral or azimuthal location may be measured, and used to control choker bars or mandrel centering devices to correct for cross-machine direction variations in thickness.

## 5 PRESSURE CONTROL

Both injection molding and extrusion utilize pressure control. For molding, cavity pressure is controlled, and for extrusion, die inlet pressure is controlled.

### 5.1 Use of Pressure Control on Extrusion

The die inlet pressure to a complex die not only determines the rate at which material exits from the die but also the cross-machine direction distribution of material. So where distribution is critical, such as for blown film or sheeting, die pressure control becomes significant. The pressure is most easily adjusted in an extruder by changing the screw speed. Increasing screw speed increases the pressure, whereas reducing screw speed reduces the pressure. Changing the screw speed slightly will give the latitude required to control die pressure. If the pressure varies wildly then the screw design, or screw speed controller, is at fault. Screw speed should never be changed dramatically to control pressure since it is also controlling solids conveying and melting as well as the pumping which generate die pressure.

## 5.2 Use of Pressure Control on Injection Molding

With injection molding the mold cavity pressure can be monitored and used to make changes in the hydraulic circuit so as to maintain the desired mold cavity pressure. Variations which are inherent from cycle to cycle such as actual times, material temperature and injection speed can be lumped together as an effect on the apparent viscosity of material which in turn affects the product quality. This apparent viscosity variation causes mold

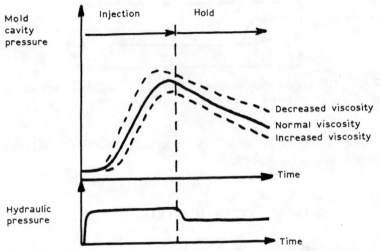

FIG. 27. Effect of apparent viscosity variation on mold cavity pressure.

cavity pressure changes to occur. The effect of apparent viscosity on mold cavity pressure is shown in Fig. 27. The hydraulic pressure, as shown, would be the same in each instance. With mold cavity pressure control, as shown in Fig. 28, the hydraulic pressure profile is adjusted to force the mold cavity pressure profile to that desired.

To compensate for the fluctuations, from cycle to cycle, in apparent viscosity the injection time is varied, along with the hold time, by changing the hydraulic pressure change over time as shown at the bottom of Fig. 28. This causes the mold cavity pressure profile to shift with time. This shift with time compensates for the effect of apparent viscosity changes as

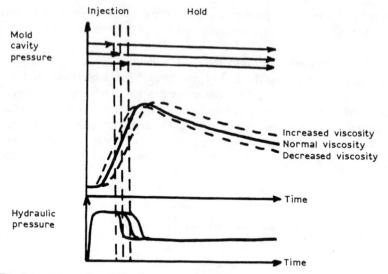

FIG. 28.    Effect of mold cavity pressure control on mold cavity pressure for a
            plastic with apparent viscosity variations.

indicated in Fig. 27. The net effect is improved control of product quality
through more consistent cycles.


## 6 OTHER TYPES OF CONTROL SYSTEMS


### 6.1 Speed Control

The speed of rotation of extruder and injection molding machine screws is
frequently open-loop with a setpoint control, but can readily be closed-
loop. Since extruders use electric motors, the armature current to the motor
is the controlled variable. Since it is linearly related to speed over small
ranges, the control algorithms are simple. It is usually advantageous to
change a number of rotational speeds on one control. For example, in a
compounding line with four feed streams and pelletizer as well as the
extruder, if we change only the extruder speed then the screw becomes
starved or flooded and the pellet length changes. So the drives are set in
ratio control to overcome this. The individual speeds can still be set relative
to the extruder speed, but if the extruder speed is adjusted all speeds change
appropriately.

## 6.2 Position Control

Position control is used in injection molding to control the forward motion of the screw during the injection stroke. It is also used in a similar yet more sophisticated fashion to control parison shape in extrusion blow molding, where a stepper motor is mounted above the mandrel. The stepper motor can drive the mandrel up or down. As it does this it opens or closes the die aperture and produces a parison thickness change. The effect is twofold. Firstly, the average parison thickness for constant-wall-thickness parison can be varied from shot to shot to reduce deviations from the setpoint. Secondly, the wall thickness of the parison can be varied along its length to allow material to be supplied where it is needed for strength but not given away where it is not required. The stepper motor is activated by the thickness measurement sensor and adjusts the position of the mandrel.

## 6.3 Time Control

In cyclic processes such as injection molding, thermoforming, compression and transfer molding, blow molding and rotational molding, the various parts of the cycle must be sequenced to ensure that the process works and is optimized. For example, in thermoforming the shoes must be heated, moved into position, the mold sealed, the vacuum drawn, and the parts cooled, ejected and removed from the mold. If this is not performed in the correct sequence then chaos will result. If the sequencing is not well controlled then the cycle time and/or the part quality will be reduced. The same holds true for all other cyclic plastics conversion processes. Timing may be controlled by clocks, but improvement can be made if these are linked or sequenced. For example, in injection molding the screw rotation is activated by the mold closure to ensure that the screw does not rotate while the mold is still open, as could happen if both were running on timers.

# 7 QUALITY CONTROL

The requirements for process control come from two sources. The first is the need to provide a workable process at economic cost. The second is the need to provide consistent product within specification. The latter need requires that measurements be made on the product, and that these measurements be within an acceptable range. An example would be that the thickness of some blown film be $0.060 \pm 0.005$ in ($1.5 \pm 0.125$ mm) at all points within the film. Of course it is impractical to measure the thickness at all points within the film so sample measurements must be made, and that these be within $\pm0.0005$ in ($\pm0.01$ mm) from the $0.060$ in ($1.5$ mm) desired value. The fact that a sample has to be taken leads to the requirement that statistics must be applied to ensure validity. In many instances, such as the one just described, it is possible to measure an attribute of the product and use this measurement to control the process. In other instances it is not. For example, in an extrusion compounding operation, the product is pellets. The attributes requiring control for these pellets may be the percentage of glass fiber, the tensile strength, the flexural modulus, the impact strength and the heat distortion temperature. These cannot be measured on-line during compounding. In fact they may not be known for many hours after the event. In instances such as these, process control limits are established on such items as screw speed, zone and die temperature. If these process control limits are adhered to, then the attributes will automatically be within specification.

Process control limits are obtained by carefully running the process and observing the controlled variables such as temperature, pressures, speeds, thicknesses, positions and times for a substantial period without making changes. When at least 30 sets of data have been obtained while an acceptable product was being produced, a statistical analysis may be made to calculate the mean and standard deviation of each variable. These data points in each variable should exhibit a normal distribution, as shown in Fig. 29. If they do not, the data are unacceptable and the test must be repeated. Once the normal distribution has been obtained, the control limits can be set. The control limits vary from process to process and must

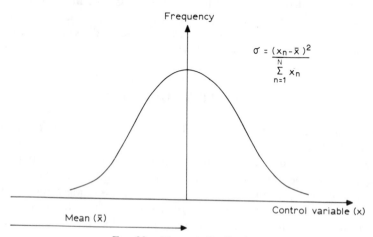

$$\sigma = \frac{(x_n - \bar{x})^2}{\sum\limits_{n=1}^{N} x_n}$$

FIG. 29.    Normal distribution.

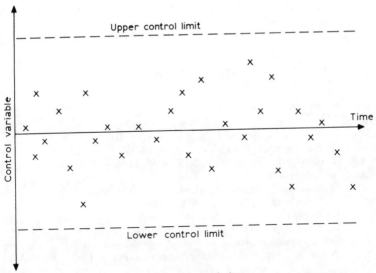

FIG. 30.    Control chart.

be set with economics in mind. Too close a control limit could be too costly to tolerate from a waste standpoint. Too lax a control limit could be too costly to tolerate from a liability standpoint. The most commonly used control limits are $\pm 3\sigma$, where $\sigma$ is the standard deviation as calculated above. At this control limit level about 5% of the product will be made while the process is outside the control limits. The control charts are set up as shown in Fig. 30. Now, as long as the controlled variables stay within the control limits, the attributes of the products will be acceptable. Operators should plot points on the control charts and changes should only be made to process setpoints in accordance with occurrences of statistical significance. Note that the specifications on the product are much wider than the acceptable attribute limits defined by the statistical process control. This is to allow for that 5% of product which is made while controlled parameters in the process are outside the control limits.

## ACKNOWLEDGEMENTS

The author would like to express his thanks to Celanese Engineering Resins Inc. for permission to publish this chapter. He would also like to thank his many associates and colleagues over the years who worked with him in the area of process control. Finally, special thanks go to Ms M. Smith, who typed this manuscript.

## BIBLIOGRAPHY

### Computer Control

GIBBONS, J. A., *Modern Plastics,* **14** (4 April 1984) 52–3.
MENGES, G., *Kunststoffe,* **30**(9/10) (Sept./Oct. 1983) 7–10.
MENGES, G., *Plastverarbeiter,* **35**(3) (Mar. 1984) 63–71.
MENGES, G., BOLDER, G., BREIL, J., ESSER, K. & MATZKE, A., *Plastverarbeiter,* **35**(9) (Sept. 1984) 48–55.
PABIOT, J., *Caoutchouc et Plastiques,* **61**(642) (June/July 1984) 51–3.
REYNE, M., *Caoutchouc et Plastiques,* **60**(635) (Nov. 1983) 83–9.
SCHEIBLBRANDUER, U., *Japan Plastics,* **34**(5) (1983) 72–4.
SCHWAB, E., *Kunststoffe,* **73**(11) (Nov. 1983) 674–9.
WORTBERG, J., *Computational analysis of polymer processing,* ed. J. R. A. Pearson & S. M. Richardson. Applied Science Publishers, London, 1983, pp. 301–35.

### Temperature Control

BURNS, R., *Plastics News (Australia)* (April 1982) 5.
BURNS, R., *Plastics News (Australia)* (Aug. 1983) 5.
FINGERLE, D. & LEUKHARDT, P., *Kunststoffe,* **66**(1) (Jan. 1976) 29–34.
FINGERLE, D., *36th ANTEC,* Society of Plastics Engrs, Apr. 1978, pp. 551–6.
HARTIG, G., *Plastverarbeiter,* **25**(10) (Oct. 1974) 612–18.
HARTIG, G., *Plastverarbeiter,* **25**(11) (Nov. 1974) 688–92.
KLEIN, R. & KLEIN, I., *Modern Plastics,* **51**(5) (May 1974) 72–4.
MEISSNER, M. & HUSTAEDT, R., *Plastverarbeiter,* **22**(4) (April 1971) 265–9.
NOREN, D., *Plastics Technology,* **28**(2) (Feb. 1982) 75–9.
POETSCHKE, H., *Kaut. u. Gummi Kunst.,* **30**(1) (Jan. 1977) 32–5.
ZEPPENFELD, H., *Plaste u. Kaut.,* **25**(5) (May 1978) 296–9.

### Thickness Control

HARRIS, H. E., *31st ANTEC,* Society of Plastics Engrs, May 1973, pp. 39–42.
HERBERTZ, T. J. M., *3rd International Conference on Instrumentation and Automation,* IFAC, May 1976, pp. 349–51.
KARAPETNITSKII, A. M., AIZENSHTEIN, M. M., VOLODIN, V. P. & AFAMAS'EVA, N. B., *Plast. Massy,* **7** (1977) 37–9.
KLEIN, R. & KLEIN, I., *Modern Plastics,* **51**(5) (May 1974) 72–4.

### Pressure Control

PARNABY, J. & WORTH, R. A., *Proc. Inst. Mech. Engrs,* **188**(25) (1974) 357–64.
PATTERSON, I. & DEKERF, T., *36th ANTEC,* Society of Plastics Engrs, April 1978, pp. 483–7.
PETTIT, G. A., *Modern Plastics,* **49**(9) (Sept. 1972) 106–7.

# Chapter 5

# POLYMERS FOR ELEVATED-TEMPERATURE USE

J. A. BRYDSON*

*Formerly Head, Department of Physical Sciences and Technology,
Polytechnic of North London, UK*

## 1 GENERAL INTRODUCTION

As the use of plastics materials becomes more widespread, the range of technical demands made upon the products becomes more wide-ranging. Some products are required to be very rigid, others highly flexible; in some cases a high level of resistance to all possible solvents is demanded, whilst in other cases water solubility may be required. Polymers may be specified as having very high electrical resistivity, whilst in other cases conductivity is a desirable attribute.

In such circumstances it is not surprising to read quite frequently of demands that a product should have a measure of heat resistance at levels which are progressively more rigorous. This results from technical developments in cars, aerospace applications, computers, domestic applications, etc.; the problem is made more acute because of trends in miniaturisation which involve polymer materials being closer to sources of heat.

Unfortunately, there is considerable misunderstanding concerning the use of terms such as 'heat resistance', and it is the purpose of the first part of this chapter to clarify terminology. Later sections will consider how 'heat resistance' may be assessed and controlled by molecular tailoring, by the use of additives and by varying the processing conditions. The chapter concludes with a brief survey of some recent introductions to the range of heat-resisting polymers.

*Present address: Brett Lodge, Brent Eleigh, Sudbury, Suffolk CO10 9NR, UK.

## 2 THE ASSESSMENT OF MAXIMUM SERVICE TEMPERATURE

A frequent need of the design engineer is for information on the maximum temperature at which a polymer may be used in a given application. In practice this will depend largely on two independent factors:

(1)  the softening behaviour of the polymer as the result of a physical change;
(2)  the thermal stability, that is the resistance to chemical change, of the polymer, particularly in air.

Additionally, further information may also be required on:

(3)  flammability;
(4)  smoke and toxic gas emission on burning.

Thus an assessment of heat resistance should involve a consideration of each of the above four components. The design engineer will, of course, normally also require information concerning properties such as stiffness, toughness, solvent resistance, and so on.

The significance of the difference between softening behaviour and thermal stability may be exemplified by considering two hypothetical polymers, X and Y. It is observed that X softens at about 120°C but has long-term thermal stability to 200°C. Polymer Y softens at 200°C but degrades 'at a measurable rate' above 90°C. Even if we allow for somewhat loose terminology, these facts indicate that polymer X could not be used much above 90°C for either long or short periods. In the case of polymer Y, short-term service might be possible up to about 170°C, but it could not be used for prolonged periods much above 70–80°C.

## 3 SOFTENING POINT AND HEAT DEFORMATION RESISTANCE

Many polymers are quite flexible, or even rubber-like, at normal room temperatures, whilst others remain rigid up to their decomposition temperature. Most thermoplastics used for engineering applications, however, are rigid at room temperature but melt at temperatures well below those at which they decompose. As the temperature is raised from room temperature to the 'melting point', there is usually a reduction in rigidity which proceeds in different ways according to the polymer used.

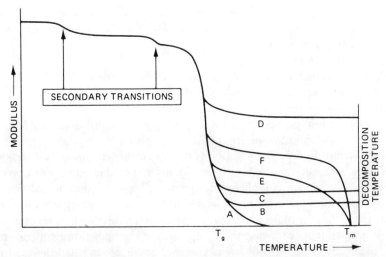

FIG. 1.   Schematic illustration of dependence of the modulus of a polymer on a variety of factors. A is an amorphous polymer of moderate molecular weight whereas B is of such a high molecular weight that entanglements inhibit flow. Similar effects are shown in C and D, where the polymer is respectively lightly and highly cross-linked. In E and F the polymer is capable of crystallisation, F being more highly crystalline than E or glass-fibre-filled. From Brydson (1982), p. 170.[2]

Figure 1 shows schematically the dependence of modulus on temperature. It will be seen from the diagram and the underlying caption that factors such as secondary transitions, the position of the glass transition temperature $(T_g)$ and, for crystalline polymers, the melting point, $T_m$, the level of crystallinity, molecular weight and presence of cross-links can all affect the shape of the curve. In turn these factors will be affected by the molecular structure of the polymer. Not shown in the figure is the influence of further factors such as the use of additives and of processing techniques, all of which are significant.

### 3.1   Control of Heat Deformation Resistance by Molecular Tailoring

The molecular factors controlling the glass transition temperature, $T_g$, have been discussed at length elsewhere by the author[1,2] and what follows here will be only a brief summary. These factors may be subdivided into the following groups.

(a)   Features associated with the behaviour of a single molecular chain which are to a first approximation independent of other molecules, e.g. chain flexibility and tacticity.

(b)  Features associated with interactions between chains, such as cohesive energy density, hydrogen bonding and side-chain spacing.
(c)  Molecular symmetry.
(d)  Copolymerisation.
(e)  Molecular weight, branching and cross-linking.
(f)  Additives such as solvents and plasticisers.

Chain flexibility is best considered by comparison with a simple polyethylene chain which in the amorphous zone, or above the crystalline melting point, is quite flexible. Such flexibility may be further enhanced by insertion of oxygen or sulphur atoms into the chain. On the other hand, insertion of ring structures such as $p$-phenylene groups will enhance the chain stiffness. Some polymers are available in which there are so many ring structures in the chain that the molecule begins to resemble a ladder with rungs and the resulting stiffness can raise the $T_g$ up to the decomposition temperature. Attachment to the main, single, chain of small groups such as methyl groups will increase the intrinsic stiffness of a chain, a feature also shown by the attachment of benzene rings. An example for the first case is made by comparing polypropylene with polyethylene, and for the second case by comparing polystyrene and polyethylene.

Attachment to the chain of polar side groups may increase inter-chain attraction, whilst hydrogen bonding can be most significant with many polymers such as the polyamides.

Although there are a number of exceptions, the glass transition temperature of a binary copolymer is usually intermediate to the glass transition temperatures of the two parent homopolymers. As a general rule, cross-linking reduces the flexibility of the polymer structure and, if sufficiently intense, will ensure that the polymer remains rigid up to its decomposition temperature.

It is not at all difficult to produce polymers which are rigid up to quite high decomposition temperatures, but this is often only achieved at the expense of processability. For this reason high levels of chain stiffness are often reduced by the insertion of occasional sulphur and oxygen atoms into the main chain to increase flexibility.

If a polymer has a sufficiently regular structure, it will normally, but not always, be capable of crystallisation. Where there are high levels of crystallinity, the polymer will remain rigid up to temperatures approaching the crystalline melting point. As a working rule of thumb, the crystalline melting point, defined as the temperature at which the last traces of crystallinity disappear, is usually about $1\frac{1}{2}$ times the glass transition temperature when expressed in Kelvin units. Thus crystalline polymers will have higher heat deformation resistance than amorphous polymers with similar glass transition temperatures. The rigidity will, however, also depend on the level of crystallisation. This difference is best seen by comparing low-density polyethylene, which has a moderate level of crystallinity, with high-density polythene with a much higher level of crystallisation.

The polymer chemist thus has considerable scope for modifying the heat deformation resistance by molecular tailoring.

### 3.2  Control of Heat Deformation Resistance by the Use of Additives

For many years it has been recognised that properties affecting the heat deformation resistance, such as $T_g$, may be affected by the addition of softeners and plasticisers. In general these reduce the heat deformation resistance. In more recent times there has been increasing use, with the so-called engineering thermoplastics, of glass fibres, and occasionally other fibrous materials, to raise heat deformation temperatures. The effect is generally most marked with polymers whose rigidity is very dependent on the level of crystallinity. One quite exceptional effect is the incorporation of glass fibre into nylon 66, which will raise the ASTM deflection temperature under load of 1·8 MPa from 75 to 245°C. Similar large increases are also shown with polybutylene terephthalate. Glass reinforcement is now becoming very common with many of the newer engineering thermoplastics, such as the polyphenylene oxides, polyphenylene sulphides, polysulphones and polyether ether ketones (Fig. 2).

### 3.3  Control of Heat Deformation Resistance by Processing

Variation of processing conditions can have significant effects on heat deformation resistance. These effects are usually most clearly shown with the crystalline polymers, since variation in moulding conditions can alter the levels of crystallinity and change the stiffness at any given temperature below the crystalline melting point. In the case of injection moulding, higher injection pressures and longer periods of time during the cooling part of the injection cycle, in which the material is between the $T_m$ and the $T_g$, will increase the level of crystallinity. Additionally, annealing at temperatures between $T_g$ and $T_m$ may cause further crystallisation to take place.

### 3.4  Measurement of Heat Deformation Resistance

Whilst fundamental data on physical changes may be obtained using

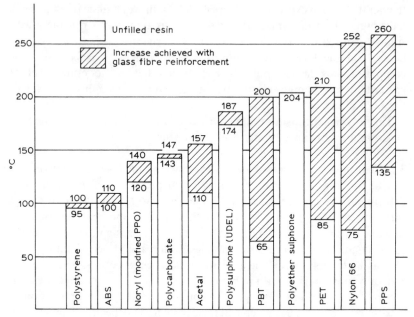

FIG. 2. Effect of glass-fibre reinforcement on heat deflection temperature under load (°C) at 1·82 MPa (264 psi) stress. Note that the effect is much greater with crystalline polymers. From Whelan & Craft (1982).[3]

differential scanning calorimetry and thermomechanical analysers, heat deformation resistance is usually assessed by measuring the temperature at which some form of modulus drops to an arbitrary value. With materials that soften over a narrow temperature range, the magnitude of this arbitrary value and the details of the test conditions in general do not have a large influence on the result obtained. Such a material is illustrated in curve A of Fig. 1.

However, with other materials that soften over a wide range or which are already somewhat flexible at or near ambient temperature, a wide range of values of 'softening point' may be obtained according to the test conditions used. Thus, for a material following curve F in Fig. 1, the use of one arbitrary level of modulus could well give a softening point just below the $T_g$, whilst another level might indicate a softening point nearer the $T_m$.

In the case of a material which followed curve D, i.e. a highly cross-linked thermosetting material, it would be possible to select a level of modulus giving a softening point just below $T_g$, whilst, at a higher arbitrary level of modulus, no softening point would be obtained below the decomposition temperature. It is clear, therefore, that the use of one-point measurements to assess heat deformation resistance can be very misleading.

In practice two tests are now used almost universally for measuring softening point and heat deformation resistance.

The first of these is the test which was known for many years as the ASTM Heat Distortion Temperature test, but which has now been adopted by other bodies, with minor modification, such as the British Standards

TABLE 1
COMPARISON OF POLYMER SOFTENING POINTS

| Polymer | Deflection temperature under load (°C) | | Vicat softening point (°C) $(10 N, 50°C h^{-1})$ |
|---|---|---|---|
| | 1·8 MPa | 0·48 MPa | |
| Polypropylene | 67 | 127 | 134 |
| Polystyrene | 95 | | 90–100 |
| ABS | 100 | | 97 |
| UPVC | 64 | 70 | 85 |
| PMMA | 97 | 106 | 114 |
| Polyacetal | 100–110 | 158–170 | 162–185 |
| Nylon 66 | 75 | 200 | 240 |
| Polycarbonate | 127–138 | 138 | 155 |
| Polysulphone | 174 | | |
| Polyethersulphone | 203 | | 226 |
| Polyphenylene sulphide | 138 | | |
| PEEK | 152 | | |
| Polyarylate (Arylef) | 175 | | |
| Polyetherimide | 200 | 212 | |

*Notes*
(1) The data are for unfilled polymer.
(2) The considerable divergence in the DTL measurements in the case of crystalline polymers.
(3) The similarities of the lower-stress DTL temperature and the Vicat softening temperature, which are both effectively measures of the temperature at which form stability is lost.
(4) The following approximate equivalents in various standards:

| | Deflection temperature | Vicat point |
|---|---|---|
| ISO | 75 (1974) | 306 (1974) |
| ASTM | D648-72 (1978) | D1525 (1976) |
| DIN | 53461 (1969) | 53460 (1976) |
| BS | 2782 Method 121 (1976) | 2782 Method 120 (1976) |

Institution and the ISO, and which is now known under such titles as the Heat Deflection Temperature under Load test or the Deflection Temperature under Stress test. In this test a specimen in the form of a bar is subjected to a three-point bending by a load which produces a maximum stress at its mid-point of either 1·8 MPa (264 psi) or 0·48 MPa (66 psi). The temperature of the specimen is raised in an oil bath heated at a controlled rate until a specified deflection of the bar occurs. The load required to produce the stress in the sample is applied to the middle point of the specimen, and this has to be calculated according to the dimensions of the bar sample. Whilst in practice one measures the temperature at which there is a specified deflection under a specified stress, in effect this means that the test measures the temperature at which a specific modulus is obtained. The use of two stress levels may or may not give very different estimates of the heat deflection temperature, according to the general modulus relationships for a polymer, as discussed in earlier paragraphs. Some typical data for heat deflection temperature at the two stress levels are given in Table 1.

The second widely-used test is that giving the Vicat softening temperature or Vicat softening point. Originally largely found in German specifications, it has also now been incorporated into ASTM, BSI and ISO specifications. In this test a standard load is applied to a circular but flat indentor which has a cross-sectional area of 1 mm² and which rests on the surface of a flat specimen. The temperature of the specimen is raised in an oil bath, which is heated at a controlled rate. When the indentor penetrates 1 mm into the material, the temperature is noted and referred to as the Vicat softening temperature. Whilst there are a number of variants of the test, the one most commonly employed is that which uses a load of 10 N and an oil bath heating rate of 50°C h⁻¹. Alternative systems, however, use heating rates of 120°C h⁻¹ and/or loads of 49 N.

## 4 HEAT STABILITY

The heat stability of a polymer is associated with its resistance to chemical changes at elevated temperatures. In some cases these changes can occur independently of the environment but, in most practical instances, the effect of oxygen, which is present in most operating environments, and moisture can cause many additional reactions.

### 4.1 Heat Destabilisation

In a large measure the heat stability of a polymer is as good as the heat stability of the weakest point of the molecule. Thus small amounts of aberrants in a polymer structure may have a destabilising effect out of all proportion to the concentration. For example, in early grades of polyacetals the rather unstable end groups provided the initiation of a chain depolymerisation reaction. Treatment of these somewhat unstable end groups by a technique known as end-capping resulted in very large improvements in stability.

Heat destabilisation can take many forms, but the most important general processes are the following.

(1) Depolymerisation.
(2) Chain scission (which may involve oxidation, ozone attack or hydrolysis).
(3) Cross-linking.
(4) Dehydrochlorination (which may then be followed by chain scission or cross-linking).

Polymer reactivity differs from the reactivity of simple molecules in two special respects. The first of these results from the presence of a number of weak links in the chains of many polymer species. These can form the site for chain scission or for some other chemical reaction. The second is due to the fact that reactive groups may occur repeatedly along a chain. These adjacent groups can react with one another to form ring products, as in the cyclisation of natural rubber and in the preparation of ladder-like polymers from polyacrylonitrile.

### 4.2 Control of Heat Stability by Molecular Structure

As indicated above, achievement of heat stability is best obtained by avoiding chemical groups and bonds of low stability, eliminating chain reactions which lead to polymer destabilisation such as unzipping, and by giving due consideration to the proximity of possible reactive groups in the polymer chain.

As a general rule, polymers containing only carbon–carbon and carbon–fluorine bonds have a high level of stability. Polymers containing only carbon–carbon and carbon–hydrogen bonds also have limited reactivity, although they may be susceptible to halogenation processes. Furthermore, where there are tertiary carbon atoms present, a site may be provided for oxidation processes. The presence of double bonds may allow reaction with such agents as oxygen, ozone, hydrogen halides and halogens to occur more easily. Some of these reactions, such as with ozone, and in some cases oxygen, can lead to chain scission and consequent breakdown of the polymer chain. Similarly, the presence of ester, amide and carbonate groups may lead to hydrolysis and, where these groups are part of the main chain, chain scission may again occur. Hydroxyl groups are also extremely reactive, and these may partake in a number of reactions.

Moderation of frequently catastrophic unzippering reactions may be achieved by the following processes.

(1) By preventing the initial formation of weak links that form the initiation site for unzippering. This will involve, amongst other things, the use of rigorously purified monomer.

(2) Deactivation of the active weak link. The end-capping of polyacetals has already been mentioned in this respect.

(3) Copolymerisation with a small amount of a second monomer which acts as an obstruction to the unzippering reaction, should this have been allowed to start. The copolymerisation of formaldehyde with ethylene oxide is an example of such an operation.

(4) The use of additives which divert the degradation reaction. A wide range of antioxidants and stabilisers function by this mechanism, which is discussed in the next section.

### 4.3 Control of Heat Stability by the Use of Additives

The use of additives to improve the thermal stability of polymers is well known and, indeed, was practised as long ago as 1870, when materials which we now recognise as antioxidants were incorporated into natural rubber. In most cases such additives work either by preferentially reacting with damaging reagents or radiation or, alternatively, by interfering with chain reactions causing degradation and diverting them into less damaging pathways.

For example, the oxidation of hydrocarbon polymers is well known to be a chain reaction with initiation, propagation and termination stages. A product of the propagation stage is an unstable hydroperoxide which can break down and initiate many further reactions. Peroxide decomposers such as phosphites and certain sulphur compounds can be most effective. More common, however, is the use of chain-breaking antioxidants which interfere with the chain propagation reaction. These are commonly phenols or amines. Both peroxide decomposers and chain-breaking antioxidants may be used together where their effect is more than additive, i.e. the system is said to be synergistic.

The stability of many chlorine-containing polymers may often be enhanced by incorporation of chemicals which act as hydrogen-chloride acceptors. Such materials are extremely important in PVC technology and, whilst there continues to be some dispute as to their exact mechanism, it appears that the removal of hydrogen chloride prevents this material, which is produced as PVC starts to decompose, from acting as an autocatalyst for further degradation.

Radiation such as from ultraviolet light can also cause degradation, and this may be partly controlled by the use of UV absorbers. Consideration of such additives is, however, somewhat outside the context of improving heat stability and will not here be considered further.

### 4.4 Control of Stability by Processing

It is no more than stating the obvious that if polymers are processed at temperatures at which destabilisation is likely to occur then some damage to the polymer will result. Not only should a polymer not be processed at a temperature above those which are necessary but, in addition, it should not be held at such undesirable temperatures for more than the minimum of time. For such reasons the plastics technologist has to ensure that flow through processing machines is positive and that there is no opportunity for material to collect in 'dead spots' or for some of the material to spend unduly long in a machine. The use of reground and reworked material clearly also has to be considered with care. Some polymers are sensitive to metals and, again, care must be taken to ensure that such sensitive polymers do not come into contact with harmful metals during the heating stage of a process.

### 4.5 Measurement of Heat Stability

For fundamental research purposes it would be common practice to study heat stability by following the chemical changes that occur during heating. On the other hand, the technologist is more concerned with monitoring the effects of elevated temperatures on processability and end-product properties.

The effect of elevated-temperature ageing will depend on the temperature of exposure and the time of exposure, and will also vary from property to property. One widely used system for comparison is that which leads to the designation of the Underwriters' Laboratories (UL) temperature index.

In order to determine the temperature index a large number of samples are oven-aged at various temperatures for up to a year. During this time samples are withdrawn and tested for such properties as flexural strength, tensile strength, impact and electrical properties. The percentage retention of a property relative to its unaged value is then plotted against time for each temperature. The time is noted at which the property has reduced 50% in value. This is somewhat arbitrarily referred to as the time to failure. A logarithmic plot of time to failure against reciprocal temperature is then made. This will be of the Arrhenius type and the line is extrapolated to yield the temperature at which failure will occur at an arbitrary time, usually 10 000 h. This is the temperature index for the property.

The temperature index will vary from property to property and with the thickness of the sample so that for a given polymer there will be a matrix of temperatures rather than a single value. The most commonly used test is that of tensile half-life, often referred to as 'mechanical-without-impact temperature index' and unless stated otherwise this is the value used in this chapter.

## 5 FLAMMABILITY

Whilst resistance to burning is a property often specified for polymers intended for use at room temperatures, it is also of particular importance for plastics being used at elevated temperatures. Resistance to burning covers such facets as flammability, smoke emission and emission of toxic gases, and this section is concerned with the first of these.

### 5.1 Flammability and Structure
As with heat deformation resistance and heat stability, it is possible to relate flammability to molecular structure. Of the various factors which have an influence, two should be specifically mentioned here.

(1)  The carbon/hydrogen ratio. As a very rough rule of thumb, it may be assumed that the higher the carbon-to-hydrogen ratio, the less will be the flammability of the material. This is shown in Table 2, which compares the carbon/hydrogen ratios and the limiting oxygen index, a measure of flammability, discussed below, for a series of polymers. If the halogen-containing polymers are excluded, the correlation coefficient for the other ten polymers is as high as 0·885.

(2)  The presence of halogens or phosphorus in the polymer structure. Table 2 also clearly shows the effect of progressively increasing the halogen level by reference to polyethylene, PVC, polyvinylidene chloride and PTFE.

Flammability may also be considerably altered by the use of additives.

TABLE 2
LIMITING OXYGEN INDEX AND CARBON/HYDROGEN RATIO
OF SOME SELECTED POLYMERS

| Polymer | LOI | C/H ratio |
|---------|-----|-----------|
| Polyacetal | 15 | 0·5 |
| Polymethyl methacrylate | 17 | 0·625 |
| Polypropylene | 17 | 0·5 |
| Polyethylene | 17 | 0·5 |
| Polystyrene | 18 | 1·0 |
| Polycarbonate | 26 | 1·14 |
| Polyarylate (Arylef) | 34 | 1·21 |
| Polyethersulphone | 34–38 | 1·50 |
| PEEK | 35 | 1·58 |
| PVC[a] | 23–43 | 0·67 |
| PPS | 44–53 | 1·5 |
| PVDC[a] | 60 | 1·0 |
| PTFE[a] | 90 | — |

[a] These polymers contain halogen atoms.

Many materials function as fire retardants, functioning by a number of different mechanisms, and include halogen-containing compounds, antimony compounds, boron-containing compounds and a number of metal oxides.

## 5.2 Flammability Tests

Two tests are widely used to characterise flammability of materials, although these will not necessarily predict how a material will perform in a real fire situation. These tests are the limiting oxygen index (LOI test) and the Underwriters' Laboratories UL94 vertical burning test.

The LOI test measures the relative flammability of plastics materials by measuring the minimum concentration of oxygen in a slowly rising stream of an oxygen–nitrogen mixture that will just support combustion. The sample is mounted in an upright position, clamped at the bottom, in a tall vertical glass column, and is ignited at the top with a propane flame. The gas ratio is adjusted according to whether or not the sample burns for more or less than three minutes. With each increase in the oxygen content the test is repeated using a fresh sample surface until the critical or minimum oxygen concentration required for three minutes' burning is found. When this critical concentration is expressed as a percentage, it is known as the limiting oxygen index. Some data for the LOI are given in Table 2.

In the Underwriters' Laboratories UL94 test, a rectangular bar is held vertically and clamped at the top. The burning behaviour and its tendency to form burning drips when exposed to a methane or natural-gas flame applied to the bottom of the specimen are noted. A record is made of whether burning drips ignite a sample of surgical cotton placed underneath the test piece, if the sample burns up to the clamp, and the duration of burning and glowing.

For the UL94 test, three ratings are of interest. They are as follows.

(1)  94V-0: specimen burns for less than 10 s after application of test flame. Cotton is not ignited. After second flame removal, glowing combustion dies within 30 s. Total flaming combustion time of less than 50 s for 10 flame applications on five specimens.

(2)  94V-1: specimen burns for less than 30 s after test flame. Cotton is not ignited. Glowing combustion dies within 60 s of second flame removal. Total flaming combustion time of less than 250 s for 10 flame applications on five specimens.

(3)  94V-2: as for 94V-1, except that there may be some flaming particles which burn briefly but which ignite the surgical cotton.

The subject of plastics flammability has been discussed in a number of excellent publications, for example by Troitzsch.[4]

## 6 SMOKE AND OTHER SECONDARY FIRE EFFECTS

Whilst each year fires cause enormous damage to property, it is a sad statistic that most deaths result from inhalation of smoke and other toxic gases. For this reason considerable efforts have been made to develop products which are less hazardous in these respects.

### 6.1 Smoke Formation

The formation of smoke depends on the incomplete combustion of a mixture of solid and liquid components resulting from pyrolysis. The pyrolysis products usually contain mixtures of aliphatic and aromatic materials, the predominance of which depends on the structure of the polymer from which they originate. In the absence of sufficient oxygen, solid smoke constituents are formed at higher temperatures, particularly in the form of soot in the flame zone. The amount of such soot is usually worse when more aromatic pyrolysis products are present. Thermoplastics which decompose primarily to aromatics in the gas phase, and thus give rise to large amounts of soot, include polystyrene, styrene–acrylonitrile copolymers and ABS. The pyrolysis of polyolefins, polyacrylates and polyacetals, polyacetals and polyamides, make only a limited contribution to smoke production. The more aromatic thermoplastic linear polymers, such as the polyethylene terephthalates and the polycarbonates, give rise to smoke-generating aromatic fragments. The more highly temperature-resistant polymers, such as polyphenylene sulphide and the polyethersulphones, tend to char and consequently evolve little smoke. Semi-ladder polymers, such as the polyimides, with their overwhelmingly cyclic structures, and thermosetting plastics, such as the phenolics, also char intensely with little smoke emission.

A variety of smoke suppressants have been developed in recent years, and their use has been reviewed.[4]

## 6.2 Smoke Tests

A large number of tests have been developed to assess smoke characteristics of burning polymers. These have been reviewed in detail[4] and will not be discussed further here. Figure 3 illustrates some data obtained by the National Bureau of Standards smoke chamber test for a variety of polymers.

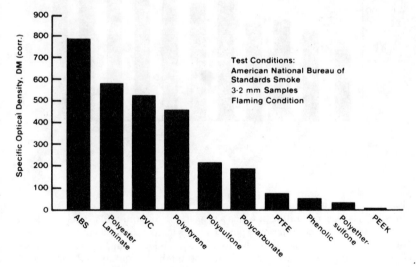

FIG. 3.   Smoke emission on burning of some plastics. (The value for polyether-imide is too small to show on the graph.) From Rigby (1983).[5]

## 7 REVIEW OF SELECTED HEAT-RESISTING POLYMERS

The range of polymers commercially available, including those for high-temperature use, has been discussed at length by the author elsewhere.[2] This review is confined to a group of aromatic thermoplastics which have glass transition temperatures (or melting points in the case of crystalline polymers) in excess of 175°C. In addition, two heat-resisting elastomers are also included for comparison.

### 7.1 General Properties

For reasons already discussed, aromatic thermoplastics generally have high

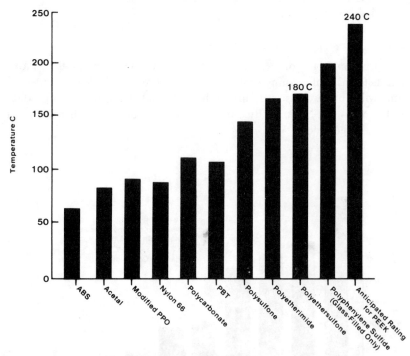

FIG. 4.    UL ratings of thermoplastics. The Underwriters' Laboratories temperature index is the temperature at which a property retains only 50% of its initial value after 10 000 hours' continuous exposure to this temperature. There are three types of index: electrical, mechanical with impact and mechanical without impact (corresponding to changes in dielectric strength, tensile strength and impact resistance respectively). These may give slightly different temperatures but are usually similar. Additives may also influence the result. From Rigby (1983).[5]

glass transition temperatures and, where they have a regular structure, a high crystalline melting point ($T_m$). Many have good thermal stability and a low flammability; a number char rather than producing a sooty flame when burning, and therefore have a wide degree of acceptability as heat-resisting polymers. The Underwriters' Laboratories have issued ratings for heat resistance, taking into account both thermal stability and heat deformation resistance, and Fig. 4 summarises the Laboratories' continuous-use temperature ratings for a selection of polymers, including some of those discussed in this section.

The aromatic thermoplastics also usually have a high degree of rigidity and strength. On the other hand, resistance to electrical tracking and similar properties are generally rather poor, and it is indeed often difficult to find data on such properties in manufacturers' published literature. Many of the aromatic thermoplastics also have a poor ultraviolet light resistance, although this may often be improved by incorporation of carbon black. Many other traditional UV absorbers cannot be used as they decompose at the high processing temperatures required with these polymers. Other properties such as toughness, stress cracking resistance and optical clarity vary from type to type. Incorporation of glass fibres and other additives such as fire retardants can also cause substantial changes in properties, and this has to be taken into account when assessing the relative creep resistance and fire properties of polymers under consideration.

Table 3 summarises selected properties of polymers discussed in this section.

### TABLE 3
PHYSICAL PROPERTIES OF HEAT-RESISTING THERMOPLASTICS

| Polymer | Glass fibre content (%) | Specific gravity | Tensile strength (MPa) | Flexural modulus (MPa) | Deflection temp. under 1·8 MPa (264 psi) load (°C) | Limiting oxygen index (%) | Izod impact strength (J/m notch) |
|---|---|---|---|---|---|---|---|
| Polysulphone (Udel) | — | 1·24 | 69 | 2 100 | 174 | 30 | 69·4 |
| Polyethersulphone | — | 1·37 | 84 | 2 570 | 203 | 34·38 | 85·4 |
| (Victrex) | 30 | 1·6 | 140 | — | 216 | — | 80·1 |
| Polyether ether ketone | — | 1·265–1·32 | 100 | 3 892 | 152 ⎫ | 24·35 | 85·4 |
| (Victrex) | 30 | — | 162 | 8 680 | >300 ⎭ | | 117·4 |
| Polyphenylene sulphide (Ryton) | 40 | 1·6 | 135 | 11 700 | >260 | 46·5 | — |
| Polyarylate (Arylef) | — | 1·21 | 65 | 1 998 | 174 | 34 | 224·2 |
| Poly-m-xylylene adipamide (Ixef) | 30 | 1·43 | 185 | 10 000 | 228 | 27·5 | — |
| Polyamideimide (Torlon) | — | 1·40 | 185 | 4 547 | 274 | 43 | 133·4 |
| Polyetherimide (Ultem) | — | 1·27 | 104 | 296 | 200 | 47 | 85·4 |
| | 30 | 1·51 | 168 | 8 957 | 210 | — | — |

## 7.2 Polysulphones and Polyethersulphones

The simplest aromatic polysulphone (formula I of Table 4) is not a practically useful material since it only melts with decomposition at about 500°C. By introducing ether groups into the chain tractable materials may be produced, and the first of these to be marketed commercially was the product introduced by Union Carbide as long ago as 1965 and now known as Udel (II).

In 1965 Minnesota Mining and Manufacturing introduced Astrel 360, to which they referred as a polyarylsulphone, probably as a consequence of its slightly higher aromaticity. In 1972 ICI introduced products to which they referred, again rather arbitrarily, as polyethersulphones (III and IV). All of these materials could be called polysulphones, polyarylsulphones or polyethersulphones but, since the distinction has been accepted by industry, it will be retained here. Of the materials used, the original commercial material, Udel, now marketed by Amoco, has the bulk of the polysulphone market, with the ICI product Victrex being used for more specialised applications demanding more stringent thermal properties. Amoco have also taken over from Union Carbide the marketing of Radel, a polysulphone of enhanced toughness.

In general, polysulphones have tended to find applications in areas for which polycarbonates are not quite able to meet the temperature requirements of specifications. Thus, whilst the $T_g$ of conventional bisphenol A polycarbonate is about 145°C, with very similar deflection temperatures under load, the $T_g$ for polysulphone is about 184°C, with a slightly lower deflection temperature. In spite of their regular structure, the polysulphones do not crystallise. It has been suggested that this may be due to the difficulty of fitting a C—S—C bond angle of 105° into a crystal lattice.

The good heat stability of the polymer has been ascribed to the high degree of resonance in the structure, giving an enhanced bond strength, and the polymer is also capable of absorbing high levels of thermal and ionising radiation without cross-linking, whilst oxidation resistance is also good. The material has an Underwriters' Laboratories continuous-use temperature of 160°C for electrical applications and applications without impact.

Polysulphone shows excellent resistance to water at elevated temperature in terms of resistance to hydrolysis, freedom from crazing and retention of impact strength. In these respects it is markedly better than both the polycarbonates and the polyetherimides. Unfilled polysulphones have a low flammability (UL94 V-0 rating at 1·5 mm) and this is further improved by the use of glass fibres and other mineral fillers. Other

**TABLE 4**

STRUCTURE AND GLASS TRANSITION TEMPERATURE OF SOME AROMATIC POLYSULPHONES

| | Type | $T_g$ (°C) | Trade name |
|---|---|---|---|
| I | [chemical structure: aromatic sulphone] | (Melts with decomposition above 500°C) | |
| II | [chemical structure with $CH_3-C-CH_3$ bisphenol group and $SO_2$, $O$] | 190 | Udel |
| III | [chemical structure with $O$, $SO_2$] | 230 | Victrex |
| IV | [chemical structure] (a) $SO_2$ / (b) $SO_2$, $O$ — (b) Predominates | 250 | Polyether-sulphone 720P |
| V | [chemical structure] (a) $SO_2$ / (b) $SO_2$, $O$ — (a) Predominates | 285 | Astrel 360 |
| VI | [chemical structure with $SO_2$, $O$] | | Radel |

outstanding properties include very good creep resistance, which is again enhanced by the incorporation of glass fibres. Important uses are in food service applications, such as microwave oven components, printed wiring board substrates, medical applications, and for fluid handling and membrane separation of gases.

The ICI polyethersulphone Victrex (III of Table 4) has a higher $T_g$ (220°C) than the Udel polysulphone and, because it also possesses good thermal stability, it has been given an Underwriters' Laboratories temperature index of 180°C. The limiting oxygen index (34–41) is slightly higher than for polysulphone (30), whilst smoke emission as assessed by the NBS smoke test (Fig. 3) is markedly less than most plastics, including polysulphone. Creep resistance is slightly better than for polysulphone, again being enhanced by the use of glass fibre fillers. As with polysulphones, the ultraviolet light resistance is poor. Whilst impact strength can be exceptional, this may be affected by temperature, moisture, processing conditions and, particularly, part geometry. The impact strength is particularly notch-sensitive, and it is important to minimise stress raisers such as notches in part designs.

The main uses for polyethersulphones have been in reusable medical equipment, membranes for use in kidney dialysis, hot-water metering systems, injection moulded, printed circuit boards and for mass-transit transport uses, where its low flammability and smoke emission characteristics are valued. The material also finds use in diverse consumer items such as hair dryers and microwave equipment.

### 7.3 Polyether Ether Ketone (PEEK) and Polyether Ketone (PEK)

The polyether ether ketones were developed by ICI, using much of the experience in chemical technology gained during the development of the polyethersulphones. They were first test marketed in 1978 and are now being sold, as with the polyethersulphones, under the trade name Victrex. The product has the repeating unit

$$-\langle\bigcirc\rangle-CO-\langle\bigcirc\rangle-O-\langle\bigcirc\rangle-O-$$

with the presence of the two ether oxygen atoms being responsible for the rather unusual nomenclature for the polymer.

PEEK has a somewhat lower $T_g$ (144°C) than any of the commercial polysulphones, but it is crystallisable with a crystalline melting point ($T_m$) of 335°C, and this often enables the material to be used at temperatures well above the $T_g$. It should, however, be recognised that there is a marked drop

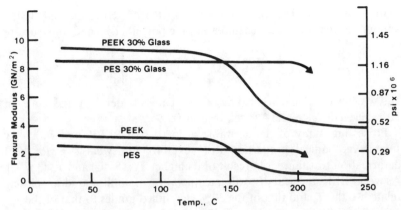

FIG. 5.    Flexural modulus versus temperature for PEEK and PES. From Rigby
(1983).[5]

in modulus at the $T_g$ (Fig. 5) and that, for unfilled materials, the ASTM deflection temperature under load is at about the same figure. However, glass-fibre-filled grades are sufficiently rigid above this temperature for heat deflection temperatures of over 300°C, i.e. approaching the $T_m$, to be reported in the literature. As with the polysulphones, the deactivated aromatic nature of the polymer leads to a high degree of oxidative stability, and it is anticipated that an Underwriters' Laboratories temperature index of about 240°C will be obtained. Although the limiting oxygen index is only 24–35, the material has a UL94 flammability rating of V-0 at 1·5 mm thickness. The smoke emission is exceptionally low (Fig. 3).

Other exceptional properties of the material are its dynamic fatigue resistance, general robustness, excellent hydrolytic stability (particularly when compared with the polyimides), very good stress cracking resistance and good radiation resistance. Applications already established include wire covering, where 'cut-through' resistance is far superior to many other heat-resisting plastics, mouldings for use in aggressive environments and abrasion- and chemical-resistant coatings.

There has been particular interest recently in carbon fibre PEEK composites marketed as Fiberite APC-2 by ICI. Such composites are 30% lighter than aluminium and in the fibre orientation direction can have tensile strengths as high as 2700 MPa. In addition to the inherent properties of PEEK summarised above, the composites also exhibit high impact toughness and a deflection temperature under load of over 300°C. One well-publicised use has been for helicopter tailplanes.

In 1986 three European companies—BASF, Hoechst and ICI—announced that they were each developing polyether ketones of structure

$$\left[ \bigcirc\!\!\!\!\!- O -\bigcirc\!\!\!\!\!- CO \right]_n$$

and which would be marketed by the companies under the trade names of Ultrapek, Hostatec and Victrex respectively.

Possessing many of the desirable properties of PEEK and with very similar mechanical properties, PEK exhibits somewhat superior heat deformation resistance as a result of its higher $T_g$ (154°C) and $T_m$ (367°C). As with PEEK, the heat distortion temperature of unfilled grades is closely related to the $T_g$ and that of the 30% glass-filled grades to that of the $T_m$.

Whilst melt temperatures of 380–420°C are necessary for injection moulding, it has been claimed that PEK is more recyclable than PEEK, a factor which could be of importance in small mouldings where the runners and sprues comprise a substantial fraction of the moulding shot.

## 7.4 Polyphenylene Sulphides

The polyphenylene sulphides were introduced by Phillips Petroleum in 1968 with full commercial production of the product, trade-named Ryton, commencing in 1973. The preparation involves the condensation of $p$-dichlorobenzene plus sodium polysulphide in a polar solvent:

$$n\,Cl\!\!-\!\!\bigcirc\!\!\!\!\!-Cl + n\,Na_2S \xrightarrow[\text{polar solvent}]{\text{heat}} \left( \bigcirc\!\!\!\!\!-S \right)_n + 2\,NaCl$$

The Celanese Corporation have recently marketed a polyphenylene sulphide under the name of Fortron.

The current materials are prepared by polymerising to a moderate degree of polymerisation of about 150, and then heating this polymer in air at a temperature below the $T_m$ to give products of substantially increased viscosity.

The polymers have a low $T_g$ of about 85°C, but the regular structure allows the polymer to crystallise with a $T_m$ of 285°C. This gives a value of $T_m/T_g$ (expressed in K) of 1·55, which is close to the rule of thumb of 1·5 quoted earlier in the chapter.

In the case of the unfilled materials, the deflection temperature, as with PEEK, is determined by the glass transition temperature but, for the glass-filled materials, the deflection temperatures quoted (>260°C) are much closer to the $T_m$. The polymer has excellent resistance to degradation in both nitrogen and air. The Underwriters' Laboratories have authorised temperature indices of 220°C for electrical applications for glass-fibre-filled grades, and 240°C for grades containing glass fibre and mineral particulate fillers. Unfilled resins will burn in a flame, but glass-filled grades have a V-0 rating by UL94 tests.

Commercially, the most important materials are fibre-filled, usually using glass fibre but occasionally carbon fibre. Some grades may also contain mineral fillers. Such reinforcement enhances mechanical properties such as stiffness and tensile strength, and furthermore helps to overcome the somewhat brittle nature of the unfilled polymer. The chemical resistance is also generally very good, with good levels of resistance to environmental stress cracking. The rather limited arc and arc tracking resistance, typical of aromatic thermoplastic polymers, has been improved by the use of glass fibre in conjunction with particulate mineral fillers. PPS is transparent to microwaves.

The polyphenylene sulphides have become established in a wide number of areas. The heat and flame resistance, coupled with good electrical insulation characteristics, has led to extensive application in electronic parts, the main application. These include connectors, sockets, coil formers, relay housings, switch relays, terminal blocks and motor housings.

A second important application area is in various mechanical parts, particularly for use in chemical processing plant, and including pump housings, impeller diffusers, pump vanes, end plates and so on. More recently, the ability to resist corrosive engine exhaust gases, ethylene glycol and petrol (gasoline) have led to increasing use in the automotive sector. One example is the replacement of aluminium by PPS in a carburettor for a small petrol engine. PPS also finds application in cooking appliances, sterilisable medical parts, dental and general laboratory equipment, and hair dryer components.

## 7.5 Polyarylates

The term polyarylate has become, again somewhat arbitrarily, associated with some highly aromatic thermoplastic polyesters. In some ways they may be considered as resulting from a natural extension of the work of Whinfield and Dickson that led to the discovery of polyethylene terephthalate in 1942 and its subsequent development as a fibre, film and moulding material. In spite of the spate of highly aromatic polymers which followed the introduction of the polycarbonates in 1959, the recently introduced polyarylates exhibit certain properties which make them of considerable potential interest.

The materials that have to date attracted the greatest attention are copolymers obtained by reacting bisphenol A with a mixture of isophthalic and terephthalic acids in the ratio 2:1:1 to give a polymer of the following general structure:

$$\left( \underset{\underset{O}{\overset{\parallel}{\underset{}{\overset{}{C}}}}{}\hspace{-0.5em}\bigcirc\hspace{-0.5em}-COO-\bigcirc-\underset{CH_3}{\overset{CH_3}{\underset{\mid}{\overset{\mid}{C}}}}-\bigcirc-O \right)$$

This polymer was first marketed by the Japanese company Unitika in 1974 and more recently under licence by Solvay as Arylef and Union Carbide as Ardel. In 1979 Bayer introduced Aromatic PolyEster (APE) with a similar structure. In the late 1980s Amoco took over from Union Carbide the marketing of Ardel, whilst Du Pont introduced further polyarylate resins under the name Arylon. The copolymers are amorphous as a result of the irregular repeating unit but the high level of aromaticity gives rise to a $T_g$ variously quoted in the range 173–194°C. Being amorphous the deflection temperature under load is, as expected, near to the $T_g$ at 174–175°C (under 1·8 MPa loading).

The fire properties are similar to the materials discussed previously in this chapter, exhibiting a Limiting Oxygen Index of 34 and a UL94 rating of V-0. However, the Underwriters' Laboratories continuous-use temperature index of 120–130°C is substantially lower than for these other materials. The polyarylates are therefore not marketed primarily for their high-temperature properties but rather as a result of other special features.

Perhaps the most interesting, both academically and commercially, is the behaviour on exposure to ultraviolet light. Whilst initially the polymers do not have good UV light resistance, they rapidly rearrange on exposure to yield o-hydroxybenzophenone structures in the polymer molecule; these are well known to have UV-stabilising effects. This reaction, known as a photo-Fries rearrangement, leads to a highly stabilised layer of polymer near the surface which in turn is able to protect the bulk of the material beneath this layer. It has been demonstrated subsequently that polyarylates may be coated on to a variety of polymer substrates such as PVC, nylon 66 and polyethylene terephthalate, and the coating allowed to rearrange on exposure to UV light to produce a UV-stable protective coating. More recently a patent has been assigned to Union Carbide for the addition of polyarylates to other polymers to improve the light stability of the latter.

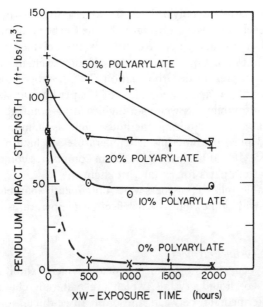

FIG. 6.    Polyarylate/PETG blend UV resistance data. (Courtesy of Union Carbide
Corp., Bound Brook, NJ.)

This is of particular interest for polyesters such as polybutylene
terephthalate and for the polycarbonates (see Fig. 6).

A second exceptional property is the very high recovery that takes place
after release of a deforming stress. In this respect the polymers are generally
superior to the polyacetals and the polycarbonates, which have been long
noted for this characteristic. This has led to considerable interest in
applications such as clips, springs, fasteners and snap-fit connectors.

The polyarylates are also of potential interest for glazing. Whilst they do
yellow somewhat due to the photo-Fries reaction, unlike the polycarbonates they do not exhibit surface haze development on exposure to
light. Thus, while not suitable for water-white glazing uses, they may be
preferred for tinted glazing purposes.

Interest for such applications and for other purposes is further
encouraged by consideration of the impact properties. Standard tests may
indicate that the polyarylates are inferior to the polycarbonates but the
situation is often reversed at elevated temperatures, with thicker sections
and where stress raisers such as notches are present. Notch sensitivity is
markedly less than for the polycarbonates.

Closely related to the polyarylates are some polyesters of a very high aromaticity level which were introduced by the Carborundum company in the 1960s. These include Ekonol, notionally the polymer of $p$-hydroxybenzoic acid, but in fact produced by the self-ester exchange of its phenyl ester in order to prevent decarboxylation. This homopolymer has a $T_m$ estimated to be 593°C and a weight loss of 1% per hour at 400°C. With a high level of insolubility as well as a high melting point, the polymer is difficult to fabricate, requiring methods such as hammering, impact moulding and pressure sintering at temperatures as high as 420°C and pressures of 35 MPa. It has been included in composites containing metal powders for use in parts for jet aircraft engines.

More tractable, related polymers were subsequently marketed under the name Ekkcel with somewhat lower deflection temperatures in the range 370–415°C.

### 7.5.1 Liquid Crystal Polymers

The $p$-hydroxybenzoic acid polymers and copolymers exhibit a number of characteristics of liquid crystals, which are materials characterised by structures intermediate between those of three-dimensional ordered crystals and disordered fluids. The molecules of these liquids may be ordered by such external forces as stress, heat flow and magnetic or electrical fields. On cooling of the melt the molecules, which are normally rod-like, retain their alignment to give exceptional tough products.

Whilst the homopolymer is very viscous in the melt and almost intractable, copolymers have been prepared which have much lower viscosities. Such a depression in viscosity may be brought about by introducing irregularities into the polymer chain by one of the following approaches.

(1) Incorporation of some flexible aliphatic links, such as may be obtained from incorporating some ethylene glycol units.
(2) Incorporation of rigid non-linear links. This may be brought about by copolymerising with monomers containing *meta*-substituted benzene rings.
(3) Frustrating chain packing by copolymerising with bisphenol-A and naphthylene-based hydroxy acids.

Commercial materials were introduced in the late 1980s by ICI as Victrex SRP (self-reinforcing polymer) and by Celanese as Vectra.

The outstanding feature of these materials is that unreinforced they show mechanical properties normally expected only from fibre-reinforced polymers. In addition, the polymers have continuous-use temperatures in excess of 200°C, are inherently flame-retardant (V-0 rating) and have exceptionally high dielectric strength. Although they are rigid materials the unreinforced polymers have impact strengths close to those of the toughest thermoplastics. The reinforced grades are much less tough. Amongst their disadvantages they have poor tracking resistance and poor abrasion resistance.

## 7.6 Poly-$m$-xylylene Adipamide

In spite of the success of the aliphatic polyamides (nylons) for uses both as fibres and plastics, the aromatic polyamides for a long time remained little more than laboratory curiosities, largely because of a number of technical problems associated with their production. Even up to the late 1970s the only significant commercial products were fibres such as Nomex, used because of its resistance to burning in protective clothing, the extremely strong Kevlar fibre and the amorphous plastics material Trogamid T, which has a $T_g$ of about 150°C.

In the late 1970s several new aromatic polyamides appeared; among these there has been particular interest in the product introduced by Mitsubishi as MXD-6 and now sold by Solvay as Ixef. It is prepared by the condensation of $m$-xylylene diamine with adipic acid:

$$H_2N-CH_2-\underset{\bigcirc}{\bigcirc}-CH_2-NH_2 + HOOC-(CH_2)_4-COOH \longrightarrow$$

$$\sim\!\!\sim NH-CH_2-\underset{\bigcirc}{\bigcirc}-CH_2-NHOC-(CH_2)_4CO\!\sim\!\!\sim$$

Unfilled polymers have the somewhat low deflection temperatures of about 96°C, close to the $T_g$ reported to be in the range 85–100°C, but the commercial materials which are glass-fibre-filled (30–50%) have deflection temperatures of 217–231°C which are close to the $T_m$ of 235–240°C. This pattern of effect of glass reinforcement is similar to that for other crystalline polymers already discussed in this chapter. Some commercial grades may contain, in addition to the glass fibre, fire retardants, impact modifiers and particulate mineral fillers.

In the absence of fire retardants the material has a limiting oxygen index of 27·5 and may burn slowly: only some grades will achieve a UL94 V-1 rating. The Underwriters' Laboratories continuous-use temperature index is also somewhat low and similar to the polyarylates with ratings of 135–140°C (electrical) and 105°C (mechanical with impact).

Initial marketing of these materials has emphasised comparisons with the aliphatic polyamides, putting particular stress on the greater stiffness, better surface finish and lower moulding shrinkage.

They have also been compared favourably with aromatic polyesters such as polybutylene terephthalate in respect of chemical resistance and polyphenylene sulphides because of the lower cost (of the polyadipamide). Early uses to be established include fishing reels and portable stereo-cassette recorders.

### 7.7 Polyimides, Polyamideimides and Polyetherimides

Polyimides are materials which contain the characteristic group

$$-N\begin{array}{c} CO- \\ \\ CO- \end{array}$$

and are thus closely related to the polyamides. However, because of the branched nature of the imide group, it is possible to produce polymers with a backbone that consists largely of ring structures. In turn, these confer rigidity to the chain and consequently a high $T_g$, or in the case of crystalline polymers a high $T_m$.

### 7.7.1 Polyimides

The first commercial materials were introduced by Du Pont in the early 1970s and were typically produced by a two-stage condensation reaction between pyromellitic dianhydride and an aromatic diamine (see formula at the top of p. 195). The polyamic acid (A) is usually soluble and mouldable but heating leads to ring closure and the product (B), which has very limited mouldability. Softening points will depend on the nature of X in the chain, but where this is, as is normally the case, aromatic the material has a very high softening point.

The zero-strength temperature of some of the polyimides is as high as 800–900°C, and in spite of fabrication difficulties polymers of the above types have found important applications for wire covering (Pyre ML), as heat-resistant films (Kapton) and machinable blocks (Vespel) as well as for composites. These are largely to be found in the aerospace and electronics industries, where there is a need for polymers with long-term high-temperature resistance in the broadest sense. The polymers have, in general, excellent oxidative stability and mechanical properties. One disadvantage is the tendency to deteriorate in hot, wet conditions.

Because they are long-established and also more akin to thermosets than

A

B

thermoplastics, these polyimides will not be considered further here and the interested reader is referred to other texts (see, for example, Refs 6–10).

### 7.7.2 Polyamideimides

In order to overcome limited processability, modified polyimides were introduced in the early 1970s. Notable amongst these were the products obtained by reaction of trimellitic anhydride with a diamine to produce a polyamideimide:

A number of grades, some of which are injection mouldable, have been made available from Amoco under the trade name Torlon since 1976. These grades may be unfilled or may contain additives such as glass fibres,

PTFE, graphite or mineral fillers; the properties are substantially altered by the choice of additive.

The unfilled polymer is notable, for an unfilled polymer, for a deflection temperature under load of about 275°C and exceptionally high tensile strengths of 172 MPa (25 000 psi), although best results are only achieved after a long so-called 'post-curing' schedule after shaping which may take several days at 260°C. The unfilled polymers have a limiting oxygen index as high as 43 with some modified grades as high as 50. The polymers have good resistance to radiation but rather poor resistance to alkalis and water. Three months' immersion can lead to a 5% increase in weight and a reduction in the deflection temperature of as much as 100°C.

### 7.7.3 Polyetherimides

In 1982 General Electric introduced yet another class of polyimide, the polyetherimide, under the trade name Ultem. The polymer has the general structure

As with the polyamideimides, the polyetherimides have exceptionally high tensile yield strengths even when not reinforced. Such unfilled polymers also have a rigidity more normally associated with a glass-filled polymer.

Furthermore, the materials have very desirable thermal characteristics. The $T_g$ of 215°C leads to a deflection temperature of about 200°C which, although lower than for the polyamideimides, is still exceptional. Because of its high thermal stability the material has been awarded a continuous-use temperature rating of 170°C (mechanical with impact) by the Underwriters' Laboratories.

Even more outstanding are the flammability characteristics. The polymer has an oxygen index of 47, amongst the highest of any thermoplastics, and a UL94 V-0 rating at 0·016″. NBS Smoke Chamber tests indicate a performance even better than for the polyethersulphones. Data on water immersion suggest excellent hydrolytic stability with a loss in tensile strength as little as about 5% after one year's immersion in water at 100°C. This is a level of performance for which the polyimides have not hitherto been noted. Chemical, radiation and UV resistance is also generally very good.

Polyetherimides may be processed on conventional injection moulding equipment although high processing temperatures are required. Similar remarks apply to extrusion and blow moulding.

The combination of properties exhibited by the polyetherimides has led to interest in automotive, domestic appliance, electrical and aerospace applications, whilst there is an emerging market for polyetherimide fibres, particularly for fire-resistant protective clothing and aircraft upholstery fabrics.

## 7.8 Polyphosphazene Rubbers

Hitherto in this chapter the polymers considered have not only possessed good heat stability and generally good fire properties but have all exhibited a high deflection temperature under load. There is, however, a class of heat-resisting polymer which have deflection temperatures way below normal ambient temperatures. These are the heat-resisting rubbers.

It is now more than 50 years since, with the introduction by Du Pont of the polychloroprene rubber Neoprene, synthetic rubbers have been available with heat resistance superior to the natural product. In the 1940s and 1950s there appeared the silicone rubbers and the early fluororubbers which pushed upward the maximum service temperatures for rubbery materials. In 1975 Du Pont marketed a fluororubber, Kalrez, to be discussed in the next section, which can withstand air oxidation to 290–315°C.

What may be regarded as the latest development is actually an extension of work undertaken some 90 years ago. It was found[11] that reaction of phosphorus pentachloride with ammonium chloride could, in appropriate circumstances, yield a totally inorganic rubber, now called poly-phosphazene:

$$PCl_5 + NH_4Cl \longrightarrow \begin{array}{c} Cl \\ | \\ +P{=}N+ \\ | \\ Cl \end{array}$$

The tendency to cross-link and also to hydrolyse on exposure to moisture delayed commercial exploitation of the polymer but eventually the polymer was used to prepare a new range of materials by substitution of the chlorine atoms in the repeat unit.[12]

Two particular derivatives to be commercially exploited were a fluorine-containing rubber (designated FZ Elastomer) and a phenol-substituted polymer (PZ Elastomer). The polymers were marketed by Firestone in the mid-1970s as PFN and APN rubbers respectively. In 1983 the Ethyl Corporation obtained exclusive rights to the Firestone patents and now market the polymers as Eypel F and Eypel A:

$$
\begin{array}{cc}
\overset{\displaystyle OCH_2CF_3}{\underset{\displaystyle OCH_2(CF_2)_xCF_2H}{\underset{|}{\overset{|}{+P{=}N)_{\overline{n}}}}}} & \overset{\displaystyle OC_6H_5}{\underset{\displaystyle OC_6H_4{-}p{-}C_2H_5}{\underset{|}{\overset{|}{+P{=}N)_{\overline{n}}}}}}
\end{array}
$$

$x = 1, 3, 5, 7, \ldots$

| | |
|---|---|
| FZ Elastomer | PZ Elastomer |
| PNF | APN |
| Eypel F | Eypel A |

The fluororubber is useful in the range $-65$ to $+175°C$. Uses have stemmed from the value of such a wide operating range combined with other desirable properties such as good damping characteristics over a broad temperature range, excellent resistance to hydrocarbon-based fluids and good flexural fatigue resistance.

The PZ rubbers, on the other hand, have a narrower service range of $-20$ to $+125°C$. They are of interest mainly for their fire properties. An oxygen index for the polymer of 28 can be raised to a value of 44 by appropriate compounding. Rather more important, the toxicity of the combustion products[13] appears to be considerably less than for many other commonly used fire-resistant elastomers. Whilst there is some smoke evolution, the compounds char on burning rather than dripping or flowing. Current interest is in wire and cable insulation and for flame-resistant closed-cell insulating foams.

## 7.9 Perfluorinated Rubbers

This chapter concludes with a brief mention of a rubber with as high a service temperature as any considered previously. This is the material marketed by Du Pont as Kalrez and which has the structure

$$
\overset{\phantom{OCF_3}}{\underset{\displaystyle OCF_3}{\underset{|}{+[(CF_2{-}CF_2)_x{-}(CF_2{-}CF)_y{-}]\!+\!cure\ site\ monomer}}}
$$

The cure site monomer is less than 2% of the total, is fluorine-containing and, as the name indicates, provides a point for cross-linking.

Vulcanisates are little affected by hydrocarbons or polar liquids, are resistant to oil-well sour gases and to a wide range of chemicals including amines, ketones and fuming nitric acid, and show continuous dry-heat resistance to 260°C and intermittently to 315°C.

The main limitations are the slow recovery from deformation, very high cost and difficulty of processing. Because of the latter, finished parts, in a limited range of shapes, are only available from Du Pont and their agents. The main use to date has been for O-rings used in severe operating conditions.

Heat resisting rubbers have been dealt with at greater length by the author elsewhere.[14]

## REFERENCES

1. BRYDSON, J. A., *Polymer Science*, ed. Jenkins, A. D., North-Holland, Amsterdam, 1972, Chapter 3.
2. BRYDSON, J. A., *Plastics Materials*, 4th edn, Butterworths, London, 1982. (5th edn, in press).
3. WHELAN, A. & CRAFT, J. L., *Brit. Plast. and Rubb.*, Nov. (1982) 29.
4. TROITZSCH, J., *Plastics Flammability Handbook* (English translation), Hanser, Munchen, 1983.
5. RIGBY, R. B., *Engineering Thermoplastics*, ed. Margolis, J. M., Marcel Dekker, New York, 1985, Chapter 9.
6. BILOW, N., LANDIS, A. L., BOSCHAN, R. H., LAWRENCE, R. L. & APONYI, T. J., *ACS Polymer Preprints*, 15(2) (1974) 537.
7. LEE, H., STOFFEY, D. & NEVILLE, K., *New Linear Polymers*, McGraw-Hill, New York, 1967.
8. RANNEY, M. W., *Polyimide Manufacture*, Noyes Data Corporation, Park Ridge, NJ, 1971.
9. ELIAS, H.-G., *New Commercial Polymers 1969–1975* (English translation), Gordon and Breach, New York, 1977.
10. BYSTRY-KING, F. A. & KING, J. J., *Engineering Thermoplastics*, ed. Margolis, J. M., Marcel Dekker, New York, Chapter 13.
11. STOKES, H. N., *Am. Chem. J.*, 17 (1897) 275.
12. ALLCOCK, H. R. & KUGEL, R. L., *J. Am. Chem. Soc.*, 87 (1965) 4216.
13. ALARIE, Y. C., LIEU, P. J. & MAGILL, J. H., *J. Combustion Toxicology*, 8 (1981) 242.
14. BRYDSON, J. A., *Rubbery Materials and their Compounds*. Elsevier Science Publishers, London, 1988.

*Chapter 6*


# THERMAL TREATMENT, PHYSICAL AGEING AND RELAXATION PROPERTIES OF AMORPHOUS POLYMERS

R. Diaz Calleja and J. L. Gomez Ribelles

*Universidad Politecnica, Escuela Tecnica Superior de*
*Ingenieros Industriales, Valencia, Spain*

## 1 INTRODUCTION

In general, the engineer is familiar with the kinds of tests, on materials, which involve an instantaneous response (deformation, polarization, etc.) to the given action (mechanical loading, electric field, etc.). For elastic materials that respond to Hooke's law, such as steel, the deformation at a given instant is directly related to the applied stress at that instant. This relationship is a result of the specific structure of this type of material. There are other groups of materials that, when stressed for a period of time in the past, still exhibit deformation as a consequence of being stressed. In other words, these materials 'remember' their previous history. A characteristic of this behaviour is that the memory 'fades' in time.

This phenomenon can affect the physical properties of polymers, particularly the amorphous types, and is shown by the temporal dependence of such material responses and at the same time is conditioned by manufacturing processes.

### 1.1 Relaxation and Creep Phenomena
The behaviour that we have just described with reference to the mechanical properties of these polymeric materials is known in the literature as viscoelastic.

### 1.1.1 Viscoelastic Behaviour

An elastic solid stores but does not lose internal energy during a mechanical test, whereas a viscous fluid dissipates but does not store energy, and therefore the intermediate behaviour is known as viscoelastic. In materials which respond to this particular type of behaviour, an instantaneous action involves a two-part response, where one is instantaneous and the other suffers damping with time. Therefore it is said that the material (a) relaxes if it has been deformed and we measure the applied force or (b) flows if it has been stressed and we measure the deformation. The mathematical description of this complex behaviour can be carried out in a linear or non-linear range, and from a physical point of view macroscopically, i.e. through modulus and compliances, or microscopically, i.e. through relaxation or retardation times.

## 1.2 Volume Relaxation

Viscoelastic mechanical relaxation has a counterpart in the dilatometric properties of these polymeric materials. In this case the tests would involve an input with a history of temperatures and the response would be the history of the change in size or volume of the sample. Normally the temperature history takes the form of a simple function, for example a single jump in temperature, and we measure the volume relaxation. This type of experiment is called 'structural recovery'. From a mathematical point of view the study of both types of relaxation (mechanical and volumetric) is quite similar. It is also interesting to note that both types of relaxation are present simultaneously in the material.[1-8]

This chapter is concerned with structural recovery processes in relation to the actual properties, and more specifically to the dynamic mechanical and dielectric characteristics, of polymeric materials.

## 1.3 The Glass Transition and the Glassy State of Polymers

The chemical structure of polymers, i.e. long chains comprising a large number of atoms linked by chemical bonds, results in physical properties different from those of low-molecular-weight substances, starting with the classification of the different states of aggregation (or phases) in which the material can be found.

### 1.3.1 Polymer Structures

The chain structure of some polymers is regular enough to fold themselves, forming a crystalline structure or matrix as a result of side group order (tacticity) or the lack of side groupings. Other types do not have this ability and therefore coil randomly, lacking a long-range order. The latter are called amorphous polymers. Polymers which possess a chain structure that only partially crystallizes and coexists with a larger or smaller quantity of amorphous phase are called semicrystalline.

The amorphous phase of polymers has no place in the usual classification of aggregation states of matter: solid, liquid and gas. An amorphous polymer is liquid-like in that it lacks a long-range order (it lacks a crystalline lattice) but differs from a liquid in that it can recover from a deformation produced by an external force when this force disappears. Two different types of behaviour can be observed in a polymer in the amorphous phase.

(1) At sufficiently low temperatures amorphous polymers show a behaviour similar to a glass—high elastic modulus, brittleness, low strain at breaking point and a low thermal expansion coefficient. This state is known as the glassy state.

(2) At a fixed temperature which is known as the glass transition temperature $T_g$ (or rather in a narrow range of temperatures around $T_g$), these characteristics suddenly change, whereby the elastic modulus becomes 100 or 1000 times smaller, the strain at breaking point is multiplied by a similar amount and the thermal expansion coefficient is doubled or tripled. This state at temperatures above $T_g$ is known as 'rubber-like'.

The glass transition temperature is usually determined by measuring the specific volume ($v$) in a test in which the temperature ($T$) is reduced at a constant rate. The glass transition is shown by a change in the slope of the curve $v(T)$. The glass transition temperature is calculated by extrapolating the curves corresponding to the 'rubbery' and 'glassy' states as shown in Fig. 1.

### 1.3.2 Molecular Mobility

All indications are that the great differences in the mechanical and thermodynamical behaviour of glass and rubber are due to differences in the segmental mobility of the molecular chain in the two states. At high temperatures a great quantity of free volume (the part of volume not occupied by the chains) exists in the material and conformational rearrangements take place easily in the chains. When the temperature decreases the material contracts; free volume and molecular mobility diminish, preventing conformational rearrangements from taking place, which leads to a closer packing of the chains. At the $T_g$ of the polymer the

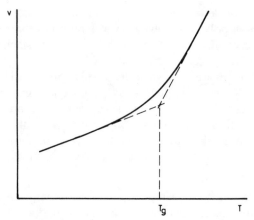

FIG. 1. Plot of specific volume ($v$) against temperature ($T$) showing the range of temperatures at which glass transition takes place and the glass transition temperature ($T_g$).

free volume is so small that the rearrangements which pack the chains closer can occur only very slowly. At temperatures below $T_g$ the specific volume measured in the polymer is larger than that which would correspond to the rubber-like state. The thermal expansion coefficient is also lower than it would be in the rubber-like state; the molecular mobility is therefore so low that the mechanical properties are similar to those of a solid. This is called the glassy state.

### 1.3.3 Physical Ageing and State of Equilibrium

Free volume and molecular mobility are very small in the glassy state but do exist. Conformational movements in the chain backbone are very slow and packing of the chains is a very slow process. This process tends at any temperature to contract the polymer to the volume of its rubber-like state at that temperature. This continuous process of contraction is known as physical ageing. The glass is in an 'out-of-equilibrium' state but undergoes a continuous process tending towards equilibrium.

The above phenomenon is detected when the specific volume is measured at a constant temperature (which can be room temperature in many polymers) after different storage times. The rate of contraction depends on temperature and on separation from equilibrium. This latter parameter depends on the thermal profile of the fabrication process (i.e. on the rate of decrease of temperature after moulding or thermal treatments undergone at temperatures below $T_g$). Other thermodynamical properties, e.g. specific enthalpy, have behaviour similar to the specific volume; the former variable (specific enthalpy) is very interesting in the study of the kinetics of physical ageing because its decrease during physical ageing is easily and accurately measured using differential scanning calorimetry.[9]

Let us consider the process which follows quenching from a temperature above to one below $T_g$ (used in many kinds of shaping processes, i.e. extrusion, injection moulding, etc.). When below $T_g$ the specific volume increases, as Fig. 2 shows, approaching equilibrium as storage time increases. The contraction rate is greater immediately after quenching, when it depends more or less linearly on the logarithm of time. The increase in density for this range of time is $c$. 0·2–0·3% for a 100-fold increase in storage time (for example from one day to three months of storage). The same increase in density could be obtained by diminishing the temperature by 10°C.

Together with the above density changes, bigger changes in molecular mobility of main chains, and therefore considerable changes in the mechanical properties of the polymer, occur. Thus, for the same increase in storage time, the mechanical relaxation time becomes one hundred times longer. Other mechanical properties such as the elastic modulus, the yield stress or the loss tangent are affected by ageing. The material can even change from a ductile behaviour to a brittle fracture with time, as in the case of polycarbonate.

This structural instability is important in some applications, such as composites with an amorphous matrix, e.g. polymer–graphite pastes that have been proposed for use as printing resistors; the stability and resistivity can be affected by the packing of the amorphous matrix (PVC for example) at room temperature.

Two problems related to the above changes in behaviour are important: firstly, the study of the kinetics of the ageing process in order to know the properties of the material after a long period of use; and secondly, the study of the influence of thermal treatments on the ageing process.

Thermal treatments can be used to place the material near enough to equilibrium so that its properties are subject to a slower rate of change and are therefore stabilized. For example, suppose that by using the appropriate thermal treatment we can bring the material to the same state that it would reach after storage for ten years. For one hundred years of use it would be necessary to change its mechanical properties to the same extent as it would suffer without thermal treatment during the first year of storage. Thermal treatments are also inherent to some uses of polymers, in aviation for example.

Therefore, in the specification of the technical characteristics of a polymeric material, it would be necessary to include (along with its conventional properties density, index of fluidity, impact strength, Young's modulus, dielectric strength, etc.) its thermal history up to the moment at which those properties were measured and the fabrication process used with special reference to the processing conditions used and the time elapsed since manufacture.

In this chapter these two problems are studied using mechanical and dielectric relaxation spectra as a basis for characterizing the molecular mobility of the polymer. A general description of the phenomenon is attempted here. A more exhaustive study would have to be carried out for each specific application of the chosen polymer.

## 2 AN EMPIRICAL STUDY OF THE STRUCTURAL RECOVERY PROCESS

In this section structural recovery is described using specific volume as a variable.

### 2.1 Structural Recovery

Firstly, structural recovery following a sudden jump in temperature, $\Delta T$, will be examined. Let it be assumed that some of the molecular groups in the polymeric material rearrange themselves immediately after $\Delta T$, whilst others need a certain amount of time to adapt to the new temperature. Let it also be supposed that the movement of each one of these latter groups contributes to a certain extent to the variations in specific volume of the system. If this were so the specific volume would follow a curve similar to the one in Fig. 2 after a sudden change in temperature, $\Delta T = T_0 - T_1$.

*2.1.1 Simplified Mathematical Treatment of Structural Recovery*
In the following, the natural logarithm of specific volume, $\ln v$, is used instead of $v$ for the sake of simplicity. The expression for the coefficient of thermal expansion in the rubbery or liquid state (in equilibrium) is

$$\alpha_1 = \frac{\partial \ln v}{\partial T}\bigg|_p = \frac{1}{v}\frac{\partial v}{\partial T}\bigg|_p \tag{1}$$

where $p$ is the pressure. Let

$$\alpha_g = \frac{\partial \ln \Delta v_g}{\partial \Delta T}\bigg|_p \tag{2}$$

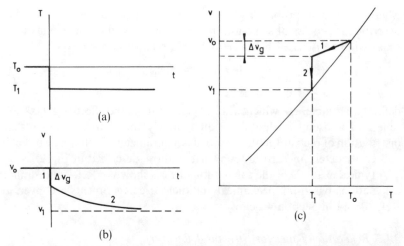

FIG. 2.   Diagram showing the change of the specific volume ($v$) with the time ($t$) (b) following the sudden fall in temperature ($T$) represented in (a). (c) The same process in a diagram of specific volume versus temperature, where $\Delta v_g$ is the increment in volume occurring immediately after the temperature change.

where $\alpha_g$ would be the value of $\alpha$ obtained in an experiment where $v$ was measured while the temperature was decreased at a much greater speed than the equilibrium approach speed of structural recovery (i.e. it is the value of $\alpha$ which is measured in the glassy state).

Structural recovery can be expressed by

$$\ln v = \ln v_0 + \alpha(t)\Delta T \tag{3}$$

so that the kinetics of structural recovery are based on the time dependence of the coefficient of thermal expansion, $\alpha$.

For a process with a single relaxation time ($\tau$), defined by the equation

$$\frac{\partial \ln v}{\partial t} = \frac{\ln v - \ln v_1}{\tau} \tag{4}$$

the time-dependent coefficient of thermal expansion would be expressed as

$$\alpha(t) = \alpha_1 + (\alpha_g - \alpha_1)\,e^{-t/\tau} \tag{5}$$

In the same way, if the process is considered to have a distribution of recovery times, then

$$\alpha(t) = \alpha_1 + (\alpha_g - \alpha_1)\int_{-\infty}^{\infty} F(\ln \tau)\,e^{-t/\tau}\,d \ln \tau \tag{6}$$

where $F(\ln \tau)$ is the probability function of the relaxation times. The need to describe the process of structural recovery by a distribution of relaxation times will be discussed below. However, a relationship such as

$$\frac{\partial \ln v}{\partial t} = \frac{\ln v - \ln v_1}{\tau_{\text{eff}}} \tag{7}$$

defines a variable, $\tau_{\text{eff}}$, which will be referred to as the effective relaxation time and is useful in describing qualitatively the characteristics of the distribution of relaxation times, since it is a parameter which can be directly and easily determined from tests such as those described in Fig. 2.

A typical example of the calculation of $\tau_{\text{eff}}$ is shown in Ref. 7 and similar calculations based on mechanical and dielectric experiments are given in Ref. 10 and in Section 4 below.

### 2.1.2 Relaxation Times for Structural Recovery
Two essential characteristics of relaxation times for structural recovery will now be considered.

#### 2.1.2.1 Dependence on temperature and structure. The relaxation times depend both on temperature and the structure of the material
Equations (6) and (7) are not linear and both $F(\ln \tau)$ and $\tau_{\text{eff}}$ depend at each instant on the value of $\ln v$. This can be seen in a diagram (Fig. 3) which represents the effective relaxation time $[\tau_{\text{eff}} = (\ln v - \ln v_1)/(\partial \ln v/\partial t)]$ (empirically determined) as a function of $\ln v - \ln v_1$ (or any other variable which indicates the separation of the system from equilibrium). Figure 3 shows qualitatively the shape of this curve for positive and negative increments in temperature.

#### 2.1.2.2 Asymmetry of structural recovery. It can be seen from Fig. 4 that there is lack of symmetry for structural recovery, i.e. the curve $v(t)$ for a fabrication process where the material is quenched from $T_1 + \Delta T$ to $T_1$ is not symmetrical with that for the process where the material is suddenly heated from $T_1 - \Delta T$ to $T_1$. This effect is a direct consequence of the dependence of the relaxation times on structure.

In both cases the process is moving towards equilibrium, which in 1 (see Fig. 4) involves a decrease in volume with time whilst in 2 there is an increase in volume with time. In this latter case the relaxation times become shorter as the system approaches equilibrium; the opposite occurs in 1. The effective relaxation times are shown in Fig. 3.

It is generally believed that the structure of the material affects all

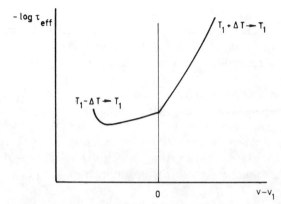

FIG. 3.   Dependence of $\tau_{\text{eff}}$ on separation from equilibrium. After quenching and during structural recovery, the volume decreases continuously. As the volume decreases so does free volume and molecular mobility, thus increasing relaxation times. On the other hand, after a sudden increase in temperature the volume is smaller than that corresponding to equilibrium, therefore the relaxation time required for structural recovery is reduced. Empirical results relating to this type of test are found in Refs 7 and 8.

relaxation times equally (presupposing the existence of a distribution of relaxation times) and in such a way that the probability function for relaxation times is constant, being displaced along the time axis as the system approaches equilibrium.[7-9]

The dependency of relaxation times on temperature can be expressed by similar functions to those used for viscoelastic or dielectric relaxation, which involve the same type of molecular rearrangements as structural recovery. This point will be discussed in detail in the following section.

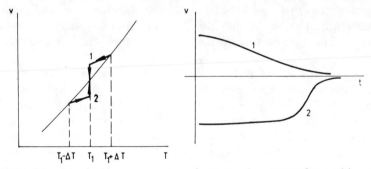

FIG. 4.   Diagram showing asymmetry of structural recovery for positive and negative temperature changes (see text).

## 2.2 Memory Effect

The tests mentioned so far comprise a sudden change in the temperature of a system in equilibrium. Another useful test involves studying changes in a system outside equilibrium when it undergoes a sudden increase in temperature. Starting with a system in equilibrium at temperature $T$, the temperature is suddenly changed to $T_1 < T_0$. After a time short enough to prevent the system reaching equilibrium at $T_1$, the temperature is suddenly increased to $T_2$. Figure 5 shows the changes in specific volume after this last increase (curve 2 in the diagram), comparing it with that which would have been observed after a single increase in temperature from $T_0$ to $T_2$ (curve 1 in the diagram).

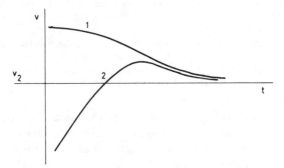

FIG. 5.    Diagram showing the memory effect.

After a double temperature jump, $T_0 \to T_1 \to T_2$, the specific volume increases initially, exceeding the equilibrium value corresponding to temperature $T_2(v_2)$ and after passing through a maximum it follows approximately the same path as it would have done had structural recovery taken place after a single jump from $T_0$ to $T_2$. This is known as the memory effect because in some way the material 'remembers' the first temperature $T_0$.

The explanation of this effect can be based on the existence of a distribution of relaxation times. Some insight can be gained by using a very simple model of relaxation time distribution in which half the molecular groups responsible for structural recovery have a relaxation time $\tau_1$ and the other half a relaxation time of $\tau_2 \gg \tau_1$, as Adachi & Kotaka have shown.[11]

To explain this effect a double tension jump experiment can be used in creep tests. The conclusion drawn[10] is also that a distribution of relaxation times is necessary for the memory effect to occur.

The analytical study of structural recovery produced by complex thermal treatment, such as more than one jump in temperature or continual variations in temperature with time, is based on Boltzmann's principle of superposition. According to this principle, the increase in specific volume from an initial value $v_0$, at time $t$, after the application of a series of $\Delta T$ at times $t_1, \ldots, t_n$ is the sum of the increases in $v$ which would have been measured at $t$ if each one of the $\Delta T$ had been applied to the material in equilibrium. Therefore

$$\ln v(t) = \ln v_0 + \sum_{i=1}^{n} \alpha(t - t_i)\Delta T_i \qquad (8)$$

or for a function $\Delta T(t)$ this becomes

$$\ln v(t) = \ln v_0 + \int_{-\infty}^{t} \alpha(t - \tau) \frac{\partial \Delta T(\tau)}{\partial \tau} \, d\tau \qquad (9)$$

In eqns (8) and (9) $\alpha(t)$ is derived from eqn (6).

The calculation of the response of a material to different thermal treatments requires a model for the distribution of relaxation times that takes into account the dependence on temperature and the structure of the material. Mention should be made of models used such as those formulated by Moynihan et al.,[9] Kovacs et al.[7] and Narayanaswamy.[12] Any model similar to those used in the empirical analysis of viscoelasticity can be used in a modified form to take into account the dependency of recovery times on volume.

## 3 INFLUENCE OF STRUCTRUAL RECOVERY ON VISCOELASTIC OR DIELECTRIC $\alpha$ RELAXATION

If the material is subjected to a mechanical stress or strain or to an electric field, a delay in the material response is produced. The following discussion makes reference to dielectric relaxation but the same would apply for viscoelastic relaxation.

At temperatures higher than the glass transition temperature ($T_g$) the system is in thermodynamic equilibrium and the immediate application of such an electric field produces a loss of equilibrium which would be followed by a series of molecular rearrangements seeking a new configuration compatible with the applied force. This process is called dielectric $\alpha$ relaxation. Although the cause which produces the loss of equilibrium is different from that which produces structural recovery, in both cases the process of approaching equilibrium is achieved by conformational rearrangements in the main chain of the polymer. The number of consecutive bonds in the main chain which take part in these rearrangements can be a function of temperature. At temperatures lower than $T_g$ these rearrangements would only be of a local order, but capable however of changing the distances between unlinked atoms. Other dielectric relaxation processes due to molecular rearrangements of small groups of the main or side chains may appear at temperatures below $T_\alpha$; they are called $\beta, \gamma, \delta, \ldots$ relaxations in decreasing order of temperature. The same nomenclature is used in mechanical relaxation.

### 3.1 Dielectric Distribution of Relaxation Times

The response to immediate application of an electric field is shown by the value of the dielectric permittivity, $\varepsilon(t)$, as a function of the time elapsed after applying the field. Hence $\varepsilon(t)$ is expressed in terms of a distribution of relaxation times as follows:

$$\varepsilon(t) = \varepsilon_U + (\varepsilon_R - \varepsilon_U) \int_{-\infty}^{\infty} F(\ln \tau)(1 - e^{-t/\tau}) \, d \ln \tau \qquad (10)$$

where $\varepsilon_U$ and $\varepsilon_R$ are respectively the unrelaxed ($t = 0$) and relaxed ($t = \infty$) permittivities, and $F(\ln \tau)$ is the probability function of the relaxation times.

Alternative tests for studying dielectric relaxation can be carried out which involve the application of an alternating field of a determined frequency and the study of the phase difference between action (electric field) and reaction (polarization). The relaxation process is characterized by complex dielectric permittivity

$$\varepsilon^* = \varepsilon' - i\varepsilon'' \qquad (11)$$

or by $\tan \delta = \varepsilon''/\varepsilon'$ as a function of the frequency of the applied field, $\varepsilon', \varepsilon''$ and $\varepsilon(\tau)$ being related by the equations

$$\varepsilon' = \int_0^{\infty} \omega\varepsilon(\tau) \sin \omega\tau \, d\tau \qquad (12)$$

$$\varepsilon'' = -\omega \int_0^{\infty} \varepsilon(\tau) \cos \omega\tau \, d\tau \qquad (13)$$

where $\omega$ is the angular frequency.

Methods of approximate calculation have been developed which, using the above equations, allow $\varepsilon'$ and $\varepsilon''$ to be calculated for a frequency range between $10^{-4}$ and $10^{-1}$ Hz from the measurement of the isothermal polarization current in a transient test.[13-15]

Dielectric relaxation becomes apparent at a specific temperature through a maximum in $\varepsilon''$ and a decrease in the value of $\varepsilon'$. An example is given in Fig. 6, where the high-frequency tail of the $\alpha$ relaxation and the $\beta$ relaxation of poly(ethyl methacrylate)(PEMA) are shown. Relaxation is characterized by a distribution of relaxation times, $F(\ln \tau)$:

$$\varepsilon'' = (\varepsilon_R - \varepsilon_U) \int_{-\infty}^{\infty} \frac{F(\ln \tau)\omega\tau \, d \ln \tau}{1 + \omega^2\tau^2} \tag{14}$$

This distribution of relaxation times can be determined by inverting the previous equations, from the results for $\varepsilon''$.[16,17] For dielectric measurements in particular, where there is a very wide frequency range available, the empirical analysis of relaxation will be more highly developed from complex dielectric permittivity than from dielectric permittivity as a function of time.

At temperatures lower than $T_g$ (in the glassy state) viscoelastic or dielectric tests are carried out from states where there is no thermodynamic equilibrium in the material. It is assumed that the kinetics of structural recovery are minimally affected by the perturbation caused by an electric field or stress.

Since the conformational movements of the main chain depend on the free volume, the relaxation time, after viscoelastic or dielectric tests, depends on the point in time at which the test is carried out. If ageing time (cf. Struik[11]) is the time elapsed from the moment in which the glassy state was formed, the viscoelastic or dielectric relaxation time will depend on the ageing time.

Since at any moment the free volume is greater than that which the system would have in equilibrium, the dielectric or viscoelastic relaxation time increases with ageing. Figure 6 shows the result of successive dielectric measurements for a sample of PEMA at $48 \cdot 6°C$ $[< T_g]$ after quenching from $100°C$ $[> T_g]$. At $48 \cdot 6°C$ the maximum value of $\varepsilon''$ corresponding to the $\alpha$ relaxation is not found within the range of frequencies measured, but the high-frequency zone (shorter relaxation times) can be seen. On the diagram the zones corresponding to the $\alpha$ and $\beta$ relaxation are marked.

It can be seen from Fig. 6 that the relaxation spectrum in the $\alpha$ zone is displaced towards lower frequencies and high relaxation times as the ageing time increases. These curves can be used to test whether the decrease in free volume affects all relaxation times equally, i.e. whether the relaxation spectrums for different ageing times are superposable by horizontal shift. If

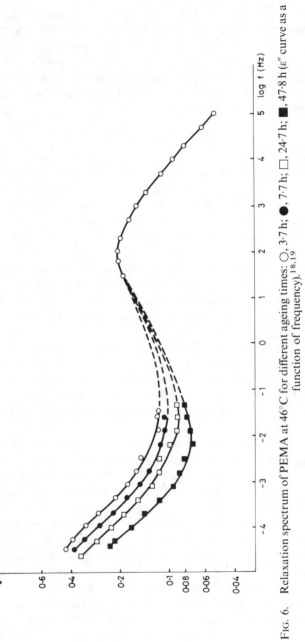

FIG. 6.  Relaxation spectrum of PEMA at 46°C for different ageing times: ○, 3·7 h; ●, 7·7 h; □, 24·7 h; ■, 47·8 h ($\varepsilon''$ curve as a function of frequency).[18,19]

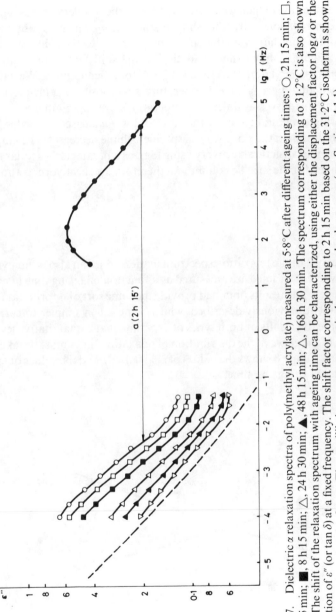

FIG. 7. Dielectric $\alpha$ relaxation spectra of poly(methyl acrylate) measured at $5.8°C$ after different ageing times: $\bigcirc$, 2 h 15 min; $\square$, 4 h 15 min; $\blacksquare$, 8 h 15 min; $\triangle$, 24 h 30 min; $\blacktriangle$, 48 h 15 min; $\triangledown$, 168 h 30 min. The spectrum corresponding to $31.2°C$ is also shown ($\bullet$). The shift of the relaxation spectrum with ageing time can be characterized, using either the displacement factor $\log a$ or the variation of $\varepsilon''$ (or $\tan\delta$) at a fixed frequency. The shift factor corresponding to 2 h 15 min based on the $31.2°C$ isotherm is shown on the figure. The equilibrium master curve is drawn as a broken curve (see Section 4.1.1).

this was the case (which we will discuss later), the recovery process could be characterized solely by the shift factor $(\log a)$ which is obtained by horizontal translation of the curves of $\varepsilon''$ versus $\log f$ at different ageing times. Figure 7 shows the calculation of the shift factor for a sample of poly(methyl acrylate) (PMA) aged at $5\cdot8°C$ for different times. We have also assumed that Ferry's time–temperature superposition principle is valid whereby the curves at different temperatures are superimposed.

During structural recovery the value of $\log a$ increases to a value $\log a_\infty$ in equilibrium in the same way as specific volume decreases. The empirical treatment of structural recovery using $\log a$ as a parameter is similar to that developed for the specific volume and therefore the effective relaxation time is defined by

$$\frac{\partial a}{\partial t} = -\frac{a - a_\infty}{\tau_{\text{eff}}} \tag{15}$$

The shift of the relaxation spectrum in the $\alpha$ zone can also be analysed by tests where $\varepsilon''$ or $\tan\delta$ are measured as a function of ageing time at a single frequency. The results obtained provide the same sort of information as the experiment previously described, whilst being much simpler to carry out. To relate the results of both types of experiment at a quantitative level, the shape of the curve of the distribution of relaxation times must be taken into account. Figure 8 shows the values of $\varepsilon''$ at $0\cdot01$ Hz and the values of $\log a$ as a function of ageing time.

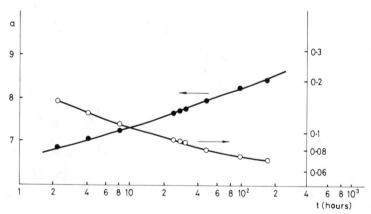

FIG. 8.    Shift factor $\log a$ and $\varepsilon''$ ($0\cdot01$ Hz) as a function of the ageing time measured in the test shown in Fig. 7.

## 4 STRUCTURAL RECOVERY FROM TRANSIENT TESTS

In the previous sections it has been seen that the glassy state of amorphous polymers is characterized by a slow but continuous process of material contraction towards equilibrium. Although from the thermodynamic viewpoint the most suitable properties for showing this phenomenon are specific volume and enthalpy, there are many other physical properties, of more technological interest, which are affected by structural recovery.

### 4.1 The Influence of Structural Recovery on Creep Tests

Struik has recently published[10] an extensive series of experimental results obtained from creep tests, aiming to show the influence of structural recovery (which he calls physical ageing) on the mechanical properties of amorphous materials over long periods of time. These tests are the mechanical equivalent of those described in Section 3 (see Fig. 7).

Some of Struik's conclusions arise directly from his experimental results and correspond with previous results obtained by Kovacs.[20] From the creep curves it is evident that physical ageing affects relaxation times as they shift towards longer times, i.e. the value of the creep compliance at a fixed creep time decreases with ageing. Figure 9 shows qualitatively the results of creep tests on samples aged at a constant temperature, after quenching from $T_1 > T_g$ to $T_2 < T_g$ for two ageing times, $t_{a_1}$ and $t_{a_2}$.

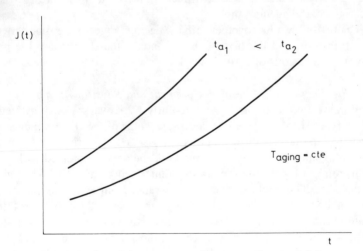

FIG. 9. Qualitative plot of the creep compliance as a function of the creep time measured after different ageing conditions (see text). cte, constant.

If the sample is heated again above $T_g$ and then requenched, essentially repetitive results are obtained for the same ageing times, which means that the process is thermoreversible, i.e. when the polymer is heated to a temperature above $T_g$ the material 'forgets' its history and so the previous ageing is erased. The construction of the master curve has been proposed for this type of test, by a horizontal shift of creep curves at different ageing times, to check if all the relaxation times are equally affected by physical ageing. Within a small dispersion the results correspond to a model of the type where

$$J(t) = J_0 \exp{(t/t_0)^m} \qquad m \sim 1/3 \qquad (16)$$

where $J(t)$ is the creep compliance at the time $t$ and $J_0$ is the reference value of the creep compliance at time $t_0$. A shift equation has been proposed on a double logarithmic shift rate which is defined as

$$\mu = \frac{\partial \ln a}{\partial \ln t_a} \qquad (17)$$

where $\ln a$ equals the value of the horizontal logarithmic shift rate along the time axis, to superimpose two curves, $J(t, t_a)$ and $\ln t_a$, the logarithmic shift rate of ageing time between both curves. According to Struik,[10] the value of $\mu$ is close to unity for all the thermoplastic materials which he studied with a characteristic treatment range of temperatures for each material, i.e. the ageing range.

The previous results, together with the ageing range for the zone between glass transition and the first of the secondary relaxations, are the most relevant conclusions of Struik's work.

### 4.1.1 Relationship of Molecular Movement and Physical Ageing

Stress relaxation tests carried out recently by Kovacs[21] on poly(methyl methacrylate) (PMMA) seem to contradict some of the above conclusions. In particular, the tests show that it is impossible to superimpose by simple horizontal shift, or by any other method, the curves obtained within the linear range of deformation at different ageing times below a given temperature. This could be due to interference from segmental rearrangement linked to physical ageing by molecular movement (probably rotation of lateral chains) causing $\beta$ relaxation. This in turn would lead to a variation in the relaxation spectrum at specific time intervals (short times) with ageing time. This raises the need to carry out measurements using different forms of stress, deformation, electric fields, etc., alongside ordered dilatometric measurements in order to interpret the different results and compare the corresponding relaxation times.

In an attempt to clarify some of the above problems we have carried out transient electric tests on PMA. This has a single secondary relaxation which occurs at very low temperatures so that the overlap with $\beta$ relaxation is small. The glass transition temperature measured by calorimetry was $8.5°C$.

Figure 7 shows the values of $\varepsilon''$ measured by subjecting the PMA sample to successive polarizations and depolarizations during structural recovery at $5.8°C$.[22] Similar tests were carried out with treatment temperatures of $3.4, 0.6$ and $-1.2°C$. Quenching took place in all cases from a temperature of around $18°C$ before the tests.

In all cases it was impossible to superimpose by horizontal shift the curves of $\varepsilon''$ versus frequency relating to different ageing times. In particular, the slope of these curves decreases slightly as ageing time increases.

The shape of the relaxation curve corresponding to equilibrium should coincide with a master curve obtained by applying the principle of time–temperature superposition at temperatures higher than $T_g$. This master curve has been drawn as a broken line in Fig. 7 in the position which would correspond to the temperature of $5.8°C$ in equilibrium, calculated by extrapolation of the position of the relaxation curves at temperatures above $T_g$ (see Ref. 22 for more details). The slope of this master curve to the right of its maximum is less than that obtained in tests with shorter ageing times. This suggests that after quenching the distribution of relaxation times is narrower, and it broadens as ageing time increases.

### 4.1.2 Relationship of Relaxation Time with Molecular Mobility

It is assumed that structural relaxation is produced by localized rearrangements of the main chain, a form of potential energy acting on each segment of the main chain. This potential energy can be considered as the sum of two parts, one of an intermolecular character, the other being intramolecular. The latter should be very similar for all the segments in the chain since it is created by the adjacent segments. On the other hand, the intermolecular energy would vary from one segment to another as a function of distance between close chains.

Hence the former is mainly responsible for the distribution of relaxation times (or of activation energies). The greater the free volume, the greater the separation between the chains, resulting in a reduction of intermolecular interaction and in a narrower distribution of relaxation times. Secondly, as chain packing increases with ageing time, the distribution of relaxation times broadens.

Due to this effect, instead of using the superposition method described in Section 3, the authors characterized the shift of the $\alpha$ relaxation spectrum by the shift on the frequency axis of an arbitrary point in the curve where $\varepsilon''$ has a definite value. For example, the value chosen was $\varepsilon'' = 0.2$ at $5.8°C$ and similar values were chosen at the other ageing temperatures. The way in which the value of $\log a$ was defined is shown in Fig. 7 and the value of $\log a_\infty$ was calculated by extrapolating the values obtained at temperatures higher than $T_g$. The values of $\log(a/a_\infty)$ are represented as a function of ageing time in Fig. 10. The shape of these curves is similar to those of dilatometric tests carried out by Kovacs,[1,7] as can be seen in particular by the curve at $5.8°C$. Effective relaxation times of the structural recovery process were calculated using eqn (15) and the results are shown in Fig. 11.

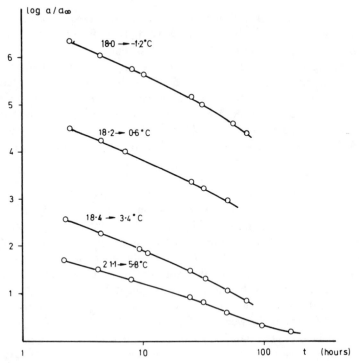

FIG. 10. The shift factor measured after temperature jumps $T_0 \rightarrow T_1$ as marked on the graph.

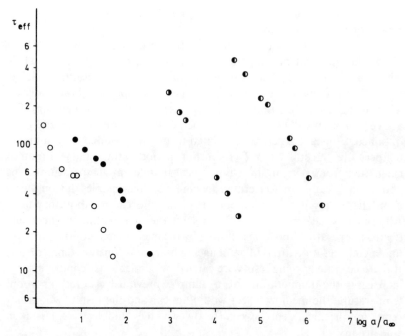

FIG. 11.    Effective relaxation times as a function of separation from equilibrium, $\log(a/a_\infty)$: ◑, $18\cdot0 \rightarrow 1\cdot2°C$; ◑, $11\cdot2 \rightarrow 0\cdot6°C$; ●, $18\cdot4 \rightarrow 3\cdot4°C$; ○, $21\cdot1 \rightarrow 5\cdot8°C$.

The double dependency of $\tau_{eff}$ on the structure of the material and on temperature can be observed.

### 4.1.3 Kinetics of Ageing Process

The kinetics of the ageing process at any temperature $T$ is determined by two parameters, $a(T, \infty)$ and $\tau_{eff}$ [see eqn (15)]. If the variable under study is different from the mechanical or dielectric relaxation spectrum, $a(T, \infty)$ would be replaced by the corresponding equilibrium value at the temperature $T$. This parameter is calculated by extrapolation of the polymer behaviour in the rubber-like state, which is easily carried out (see, for example, Refs 7, 10 and 26).

The effective relaxation time depends on the temperature and on departure from equilibrium [or on the value of the parameter $a(T, t_a)$ in this case]. The equation

$$\log \tau_{eff} = \log A(T) - B \log(a/a_\infty)$$

where $A$ is a function of temperature and $B$ is a constant, is suitable to represent this double dependence.

Parameter $B$, which can be calculated from the slope of the $\log \tau_{\text{eff}}(t_a)$, is almost independent of temperature, as can be deduced from Fig. 11, in which $\log \tau_{\text{eff}}(t)$ curves are almost parallel for the four temperatures studied. Suitable tests could be devised to determine the properties of the materials after storage for a long time at a low temperature (long-term ageing). The above properties will be determined when values for constants $A$ and $B$ are obtained. Constant $B$ can be calculated from a test conducted at a higher temperature (but still below $T_g$) for a short period of time, and $A(T)$ can be calculated from several short-time tests at different temperatures. The above can be called an accelerated ageing test, which enables the long-term properties of a polymer to be obtained from short-term laboratory tests. All tests must be carried out with the same thermal treatment: quenching from a temperature above $T_g$ followed by the measurement of the property under consideration after different ageing times at a constant temperature. If the accelerated ageing tests were carried out under other conditions, e.g. increasing the temperature of a sample previously stored at room temperature, the memory effect would appear and the results would not represent the ageing process correctly, which would lead to mistaken conclusions.

## 5 STRUCTURAL RECOVERY FROM ALTERNATIVE TESTS

In addition to the experimental contradiction referred to in the previous section, it is obvious that creep or stress relaxation tests do not allow direct observation of secondary relaxation and therefore the effect of physical ageing cannot be observed.

In these circumstances dynamic techniques are especially useful as they allow very accurate investigation of the specific secondary relaxation phenomena as related not only to the morphology of the material, but also to thermal treatments and in particular to physical ageing.

### 5.1 Dynamic Tests
The usual physical parameters measured by traditional dynamic techniques are the dynamic moduli or compliance in mechanical tests and dielectric permittivity and/or the tangent of the loss angle in both types of tests. This last parameter is a dimensionless quantity which is independent of sample geometry and shows minimal variations irrespective of the test

used[23] and it is therefore ideal to describe the influence of ageing. As discussed in Section 3, the shift of the $\alpha$ relaxation spectrum due to physical ageing can be analysed in terms of the tangent loss through tests at constant frequencies. This type of test is much more convenient when working with mechanical techniques in which the experimental frequency interval is not usually longer than two or three decades.

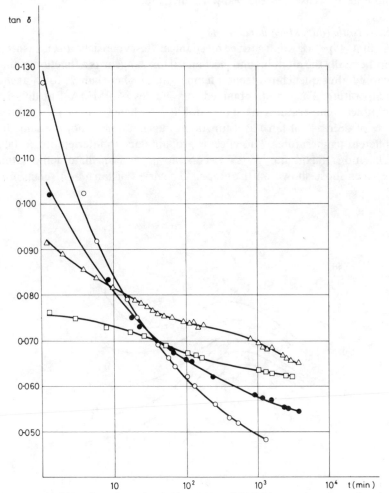

FIG. 12.   Values of tan $\delta$ mechanical at 1 Hz of PMMA as a function of ageing times at different temperatures. The samples were quenched from 150°C to ageing temperature. ○, 77°C; ●, 59·5°C; △, 46°C; □, 28°C.

The loss tangent decreases as the ageing time increases in a similar way to that shown in Fig. 8 for the dielectric loss permittivity, in a process which is qualitatively similar to a decrease in specific volume with ageing time. This is due to a decrease in molecular mobility brought about by a decrease in free volume linked to physical ageing so that the relaxation spectrum shifts towards longer times. The imaginary parts of the complex compliances and modulus decrease and the real parts increase.

### 5.1.1 Isothermal Mechanical Tests

A first type of alternative mechanical test consists in measuring isothermally the modulus and mechanical loss tangent as a function of time elapsed after quenching, from a temperature higher than $T_g$ to an ageing temperature. The results obtained for samples of PMMA[24] at different treatment temperatures are shown in Fig. 12, where it can be seen that the rate of decrease of $\tan \delta$ as a function of ageing time is not the same for different temperatures. This effect is probably due to interference at the $\beta$ relaxation maximum, which occurs at an intermediate temperature between those shown on the graph. The curves of $\tan \delta$ as a function of

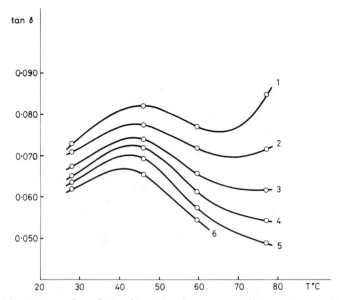

FIG. 13.    Curves of $\tan \delta$ as a function of temperature for constant ageing times obtained from the results shown in Fig. 12. 1, 10 min; 2, 30 min; 3, 100 min; 4, 300 min; 5, 1000 min; 6, 3000 min.

temperature at constant ageing time are obtained by interpolation in Fig. 12 and are shown in Fig. 13.

### 5.1.2 Double Temperature Jump Tests

In double temperature jump tests as described in Section 2.2, where the sample is quenched from $T_0 > T_g$ to $T_1 < T_g$ and after a certain interval of time a new positive temperature jump to $T_2$ is applied to the sample, so that $T_1 < T_2 < T_0$ and $\tan \delta(T_2) > \tan \delta(T_1)$, the memory effect can be observed.

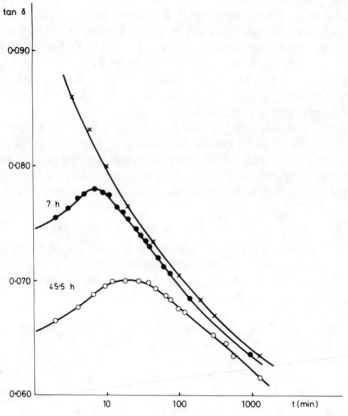

FIG. 14. Results of double temperature jump tests on a sample of poly(methyl methacrylate) (PMMA), which was quenched from 150 to 28°C and this temperature was maintained for distinct intervals of time, i.e. 7 h (●) and 45·5 h (○) respectively for each curve of tan δ. The temperature was then increased to 46°C and results of these tests were compared with those obtained after quenching from 150 to 46°C (×).

The values of tan $\delta$ increase at first during an interval of time which depends on the ageing time prior to $T_1$, finally exceeding the value of tan $\delta$ corresponding to equilibrium at temperature $T_2$, then they go through a maximum and approach values which a sample aged directly to temperature $T_2$ (Fig. 14) would have.

### 5.1.3 Non-isothermal Mechanical Tests

A second type of test consists in quenching the samples to the chosen temperature for ageing. After different periods of time, the ageing process is suddenly stopped by immersion of the samples in liquid nitrogen. The dynamic modulus and tan $\delta$ are then measured as a function of temperature at a constant rate of temperature increase. The results for tan $\delta$ of PMMA at 1 Hz aged to $59.5 \pm 0.5°C$ during different intervals can be seen in Fig. 15. The graph shows the effect which structural recovery has on the value of tan $\delta$, which decreases with the ageing time in the range of temperatures between approximately 20 and 80°C.

In this type of test the effect of ageing may be overlapped with that of the thermal treatment, which implies the continuous increase of temperature during the measurements; thus, at temperatures higher than the ageing temperature, the previous ageing effect disappears creating anomalous relaxation peaks, probably as a result of the memory effect (see Fig. 16).

### 5.1.4 Dielectric Tests

Figure 17 shows the results of a series of dielectric tests on samples of various polymers which have been systematically given two types of thermal treatment: (1) quenching to room temperature from a temperature higher than $T_g$; (2) slow quenching (usually $5°C\,h^{-1}$) from $T > T_g$ to room temperature.[25]

This type of test is seldom used for physical ageing since it is not isothermal; however, it does allow useful conclusions to be reached. On the whole, the values of tan $\delta$ in the range analysed are higher for the sample which was quenched than for the sample slowly cooled. For PVC indications of a new relaxation occur; this matter will be discussed in Section 6 in terms of mechanical tests. The range of temperatures in which this phenomenon occurs is variable according to the type of polymer used and the frequency at which it is measured. When considering PMMA this phenomenon cannot be related to the position of $\beta$ relaxation as there is no such range at high frequencies (i.e. producing an overlap of tan $\delta$ curves for the samples subjected to both types of treatment).

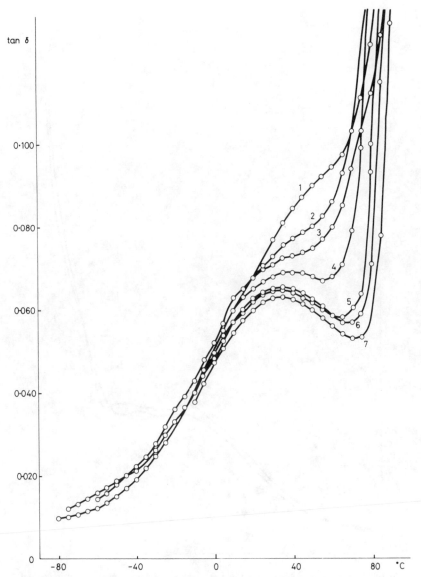

FIG. 15.   Values of tan $\delta$ as a function of temperature measured as the temperature of the sample was increased at $3°C\,min^{-1}$ from $-80$ to $140°C$ (only results between $-80$ and $100°C$ are represented in the graph in order to show clearly the effect of ageing). The ageing temperature was $59\cdot5 \pm 0\cdot5°C$. (1) Quenching; (2) 10 min; (3) 1 h; (4) 17 h; (5) 46 h; (6) 168 h; (7) 624 h.

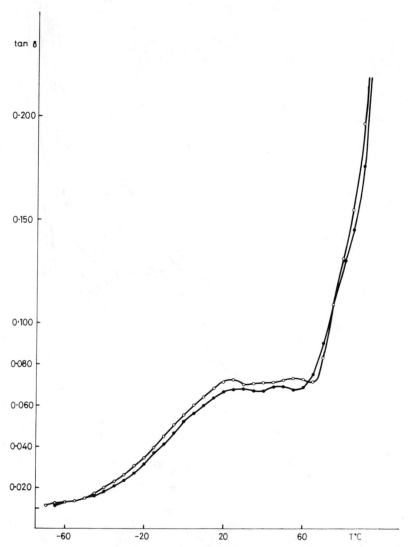

FIG. 16.   Values of tan $\delta$ measured at 1 Hz and at the rate of 3°C min$^{-1}$ from $-70$ to 100°C for a sample of PMMA aged at 46°C for (●) 22 h, (○) 65 h and then quenched to $-70$°C before measuring.

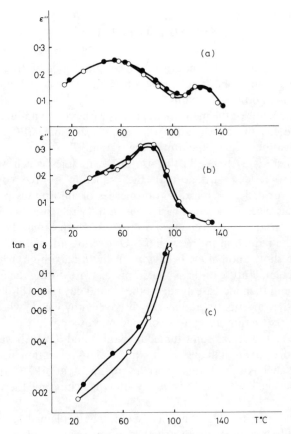

FIG. 17. Values of $\varepsilon''$ or $\tan \delta$ as a function of temperature for samples of PMMA [poly(methyl methacrylate)] (a), PMEA [poly(ethyl methacrylate)] (b), PHEMA [poly(hydroxyethyl methacrylate)] (c) quenched from a temperature higher than $T_g$ to room temperature (●) or cooled at $5°C\,h^{-1}$ within the same temperature range (○).

We have attributed this to the fact that a decrease in free volume caused by slow cooling particularly affects the main chain segments and therefore the range of low-temperature relaxation, but it hardly affects $\beta$ relaxation which has a qualitatively very different molecular origin. For this reason, in PMMA where the magnitude of $\beta$ relaxation is higher than $\alpha$ and where it overlaps at high frequencies, the ageing effect can only just be observed on the $\varepsilon''$ curves.

## 6 POSSIBLE CHANGES IN LOCAL ARRANGEMENTS DURING THERMAL TREATMENT

It has been suggested[26,27] that there is a certain degree of ordering in the main chains in amorphous polymers, consisting in the alignment of chains to form small nodules (domains). The movements of the main chain groups within these nodules could be responsible for the $\beta$ relaxation of the polycarbonate (PC).[28] During thermal treatment at temperatures lower than $T_g$ as well as structural recovery there is a change in the size of the nodules which also affects (although to a lesser extent than the reduction in free volume) the viscoelastic or dielectric relaxation properties of amorphous polymers.

Geil[29] has discovered small variations in the $\gamma$ relaxation of PC (along with other variations in mechanical properties) which cannot be explained solely by the reduction in free volume. Our results for the PC treated at 120°C also show anomalous behaviour in the $\beta$ maximum, whereby the height and temperature of the $\beta$ relaxation decrease during the first day of treatment and then increase again on the second day (Fig. 18). This can be related to the variation in nodule size discovered by Geil[29] when viewing the polymer structure using an electronic microscope.

PVC also has a relaxation situated between $\alpha$ and $\beta$ (which we call $\beta'$)[30] which is more apparent in quenched samples. This relaxation also presents anomalous behaviour during thermal treatment at 61·7°C. Its height increases between one hour and one day after quenching and then decreases (Fig. 19).

As in the case for PC, this effect cannot be explained solely by the decrease in free volume during structural recovery and could be related to another structural change in the material similar to that which occurs in PC.

## 7 INFLUENCE OF STRUCTURAL RECOVERY ON SECONDARY RELAXATIONS

The molecular groups which produce the secondary relaxations are in general smaller than those which produce $\alpha$ relaxation and the potential barriers which oppose their movements are mainly intramolecular.[31] This suggests that structural recovery has little influence on these relaxations. The experimental justification of this meets with serious difficulties in the

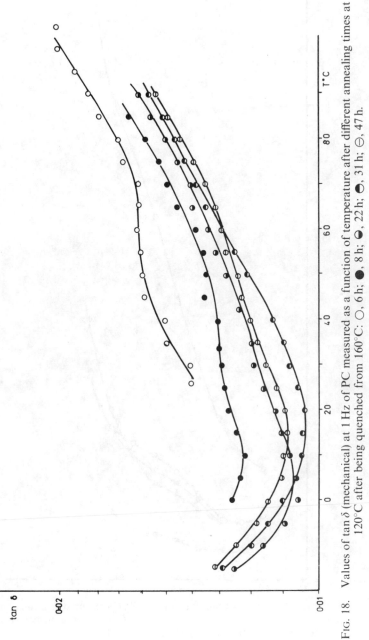

FIG. 18. Values of tan δ (mechanical) at 1 Hz of PC measured as a function of temperature after different annealing times at 120°C after being quenched from 160°C: ○, 6 h; ●, 8 h; ◑, 22 h; ◐, 31 h; ⊖, 47 h.

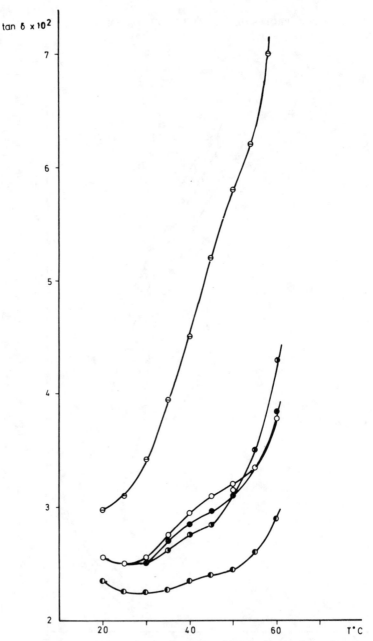

FIG. 19. Values of tan δ (mechanical) at 1 Hz of PVC measured as a function of temperature after different annealing times at 61·7°C after being quenched from 100°C: ◑, 1·25 h; ●, 5 h; ○, 23 h; ◐, 4 days, quenched sample.

case of relaxations close to glass transition such as is the case in PMMA $\beta$ relaxation.

In the range of temperatures at which this relaxation is produced, there is overlapping with the low-temperature zone of the $\alpha$ relaxation and it is not clear which is responsible for the variations observed during ageing. Thus, in dynamic mechanical tests (Fig. 13), a reduction in height can be observed with treatment time which does not occur in dielectric measurements in which the $\alpha$ maximum has less intensity than $\beta$ (Fig. 16).[25,32] This is similar to results obtained for PEMA (Fig. 6).

In secondary relaxations which occur at lower temperatures some variations have been discovered by thermal treatment at temperatures lower than $T_g$,[32,33] but the variations are small and the results available are not sufficient for a theory to be developed as yet. More experimental work is in progress.

# 8 CONCLUDING REMARKS ON PHYSICAL AGEING

This work presents an outline of the most important physical facts related to structural recovery. At first sight it may appear that it is a microphenomenon affecting the properties of the material only at a molecular level. However, the densification process associated with physical ageing is of outstanding technological interest. It is well known that for economic reasons the processing conditions for the amorphous polymers are more severe than for semicrystalline polymers in which it is necessary to ensure immediate dimensional stability after injection moulding. The quenching which finishes the processing of amorphous polymers sets on long-time processes that affect continuously the wearing properties of the material during many years. For example, there is a problem of dimensional stability in precision pieces; also cracking arises in the stressed parts of polymeric material in complicated mechanisms. Moreover, although the decrease in the specific volume during ageing is small, we have observed important changes in macroscopic properties, as for example the dynamic mechanical modulus $E^*$. For a sample of PMMA aged at 59·5°C this modulus increases by about 10% between the first 10 min and 1 h after quenching. The decrease of the loss tangent is still larger. This is only an example of the importance of knowing and controlling the thermal history of polymers in order to predict changes in their properties with time.

# REFERENCES

1. KOVACS, A., *J. Polym. Sci.*, **30** (1958) 131.
2. SAITO, N., KOANO, K., IWAYANAGI, S. & HIDESHIMA, T., *Solid State Physics*, Vol. 14, ed. F. Seitz and D. Turnbull. Academic Press, New York, 1963.
3. NEKI, K. & GEIL, P. H., *J. Macromol. Sci.—Phys.*, **B8** (1974) 295.
4. PETRIE, S. E. B., *J. Macromol. Sci.—Phys.*, **B12** (1976) 225.
5. AREF-AZAR, A., BIDDLESTONE, F., HAY, J. N. & HAWARD, R. N., *Polymer*, **24** (1983) 1245.
6. PETRIE, S. E. B., in *Polymeric Materials*. American Society for Metals, Metals Park, Ohio, 1975.
7. KOVACS, A. J., AKLONIS, J. J., HUTCHINSON, J. M. & RAMOS, A. R., *J. Polym. Sci.*, *Polym. Phys. Ed.*, **17** (1979) 1097.
8. ROBERTSON, R. E., *Ann. N.Y. Acad. Sci.*, **371** (1981) 21.
9. MOYNIHAN, C. T. *et al.*, *Ann. N.Y. Acad. Sci.*, **279** (1976) 15.
11. ADACHI, K. & KOTAKA, T., *Polym. J.*, **14** (1982) 959.
10. STRUIK, L. C. E., *Physical Aging in Amorphous Polymers and Other Materials*. Elsevier, Amsterdam, 1978.
12. NARAYANASWAMY, O. S., *J. Am. Ceram. Soc.*, **54** (1971) 491.
13. HAMON, B. V., *Proc. Inst. Elec. Eng.*, **99** (PEIV Monograph) (1952) 27.
14. BRATHER, A., *Colloid. Polym. Sci.*, **257** (1979) 785.
15. GOMEZ RIBELLES, J. L. & DIAZ CALLEJA, R., *J. Polym. Sci.*, *Polym. Phys. Ed.*, **23** (1985) 1297.
16. FERRY, J. D., *Viscoelastic Properties of Polymers*. Wiley, New York, 1970.
17. SCHWARZL, F. R. & STRUIK, L. C. E., *Advan. Mol. Relaxation Processes*, **1** (1967–68) 201.
18. GOMEZ RIBELLES, J. L. & DIAZ CALLEJA, R. (1985), unpublished results.
19. GOMEZ RIBELLES, J. L. & DIAZ CALLEJA, R., *Anales de Fisica*, **81** (1985) 104.
20. KOVACS, A. J., ScD Thesis, University of Paris, 1955.
21. MCKENNA, G. B. & KOVACS, A. J., *Polymer Preprints*, **24** (1983) 100.
22. GOMEZ RIBELLES, J. L. & DIAZ CALLEJA, R., *Polym. Bull.*, **14** (1985) 45.
23. HEIJBOER, J., Dr Ciencias Naturales thesis, University of Leiden, 1972.
24. GOMEZ RIBELLES, J. L. & DIAZ CALLEJA, R. XX Reunion Bienal de la RSEQ, Comunicacion 2.1.
25. GOMEZ RIBELLES, J. L. & DIAZ CALLEJA, R., *Polym. Eng. Sci.*, **24** (1984) 1202.
26. GEIL, P. H., in *Polymeric Materials*. American Society for Metals, Ohio, 1973, Chapter 3.
27. YEH, G. S. Y., *Crit. Rev. Macromol. Sci.*, **1** (1972) 173.
28. SACHER, E., *J. Macromol. Sci.—Phys.*, **B10** (1974) 319.
29. NAKI, K. & GEIL, P. H., *J. Macromol. Sci.—Phys.*, **B8** (1974) 295.
30. DIAZ CALLEJA, R. & GOMEZ RIBELLES, J. L., *Polym. Eng. Sci.*, **22** (1982) 845.
31. HEIJBOER, J., *Ann. N.Y. Acad. Sci.*, **279** (1976) 104.
32. GUERDOUX, L. & MARCHAL, E., *Polymer*, **22** (1981) 1199.
33. ITO, E., TAJIMA, K. & KOBAYASHI, Y., *Polymer*, **24** (1983) 877.

## Chapter 7

## DIGITAL HYDRAULICS

S. Skinner

*Vickers Systems Ltd, Havant, Hampshire, UK*

and

H. Briem

*Vickers Systems GmbH, Bad Homburg, West Germany*

## 1 INTRODUCTION

Poppet-type valves have been used in hydraulic systems for many years, but since 1970 their use has been extended and now covers virtually the full range of control valve functions. Generally referred to as *cartridge valves* or logic elements, their use is aimed at providing a more efficient and compact control system. As industry becomes more energy-conscious, cartridge valves, together with power-saving pump controls, can show significant benefits over traditionally accepted levels of hydraulic system efficiency.

Cartridge valves provide an alternative to gasket- or pipe-mounted valves, and in many applications can show advantages such as:

(a) system design flexibility;
(b) lower installed cost;
(c) smaller package size;
(d) better performance and control;
(e) improved reliability;
(f) higher pressure capability;
(g) more efficient operation;
(h) lower internal and external leakage;
(i) more contamination tolerance;
(j) faster machine cycle times.

It must be remembered, however, that cartridge valves offer an alternative rather than a replacement for conventional sliding spool valves. Engineers faced with a design problem should consider all the possibilities before deciding on a particular solution. Cartridge valve systems will become cost-effective if some of the following design parameters are significant:

- (a) high flow rates;
- (b) high working pressure;
- (c) small envelope size;
- (d) complex circuitry;
- (e) series production;
- (f) fast response;
- (g) low internal and external leakage;
- (h) good stability;
- (i) low noise;
- (j) reliability;
- (k) low contamination sensitivity.

## 2 BASIC PRINCIPLES

### 2.1 Operating Principles

A cartridge valve employs basically a seated poppet, as shown in Fig. 1, with two main flow ports, A and B, and a pilot control port, Ap. The poppet

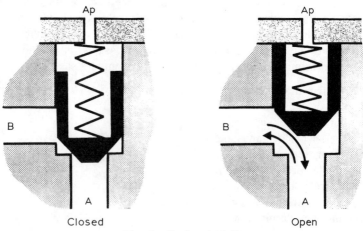

Closed                                          Open

FIG. 1.    Basic principle.

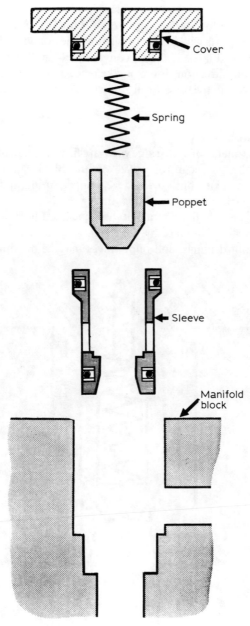

FIG. 2. Cartridge valve components.

can be in one of two positions, i.e. closed, when the A and B ports are blocked, or open, allowing flow from A to B (or B to A). The pressure in each of the three ports will determine whether the poppet is open or closed, as will be described later. Unlike a conventional check valve which only allows flow in one direction, flow can take place in both directions through a cartridge valve and the pressure in *either* the A or B port will tend to open the poppet.

## 2.2 Main Components

The main components of a cartridge valve are fitted inside a manifold block leaving only the pilot control valves on the block surface. The basic components (insert kit) consist of a sleeve, poppet, spring, and the necessary 'O' rings and back-up rings as shown in Fig. 2.

The insert assembly fits into a suitably machined hole in the manifold block and is retained in place by the cover. The cover plate contains the appropriate pilot drillings, and pilot valves may be mounted on the top

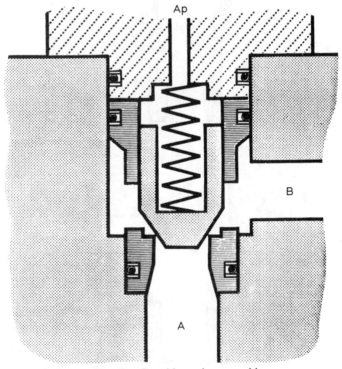

FIG. 3.   Cartridge valve assembly.

surface of it as required. The two main flow ports are also machined in the block, one at the bottom of the recess and the other in the side.

Although the components of the insert kit will vary depending upon the valve function, the dimensions of the recess in the manifold remain constant for any particular size of cartridge. Cartridge valve sizes are specified by referring to the outside diameter of the poppet; for example, NG 16 = 16-mm poppet, NG 25 = 25-mm poppet, etc.

The cartridge insert (Fig. 3) can be considered as the main stage of a two-stage valve. System flow takes place through the A and B ports while the pilot valves determine whether the valve is open or closed by controlling the pressure on the Ap port.

## 2.3 Leakage

A significant advantage of cartridge valves is that when the poppet is closed virtually zero leakage exists between the A and B ports, since the poppet and seat provide a positive seal. Leakage can exist, however, between the B

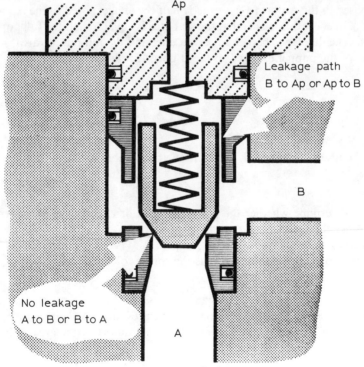

FIG. 4. Insert leakage.

port and Ap port (or vice versa) as this relies purely on the sliding fit between the poppet and sleeve (Fig. 4).

When designing cartridge valve systems, thought must also be given to the effect of leakage through pilot valves, as will be discussed later.

## 3 POPPET TYPES

In order to achieve different control functions, several types of cartridge insert are required which will be described in detail.

### 3.1 The 2:1 Ratio Poppet
This is the most common insert and is used for check and directional valve functions.

#### 3.1.1 Poppet Dimensions
The dimensions are such that in the closed position the area of the poppet exposed to the A port ($A_A$) is exactly half the area exposed to the Ap port ($A_{AP}$). In effect this means that the seat area is equal to half the full poppet area. The difference between the seat area and the full poppet area is the area of the poppet exposed to the B port. It follows therefore that

$$A_A = A_B = \tfrac{1}{2}A_{AP}$$

or

$$A_{AP} = 2A_A = 2A_B$$

(see Fig. 5).

The insert is defined by the ratio of the Ap area to the A area, i.e.

$$A_{AP}:A_A = 2:1$$

To determine whether the poppet is open or closed, it is necessary to consider the forces acting on the poppet.

The force tending to close the poppet ($F_{close}$) is the pressure at Ap ($P_{AP}$) multiplied by the area of Ap plus the spring force ($F_s$), i.e.

$$F_{close} = P_{AP} \cdot A_{AP} + F_s$$

The force tending to open the poppet is the pressure at A ($P_A$) multiplied by the area of A, plus the pressure at B ($P_B$) multiplied by the area of B, i.e.

$$F_{open} = P_A \cdot A_A + P_B \cdot A_B$$

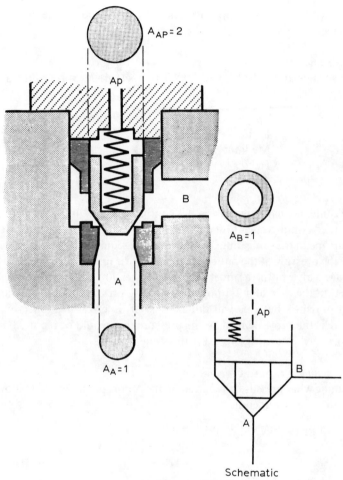

Fig. 5. The 2:1 ratio poppet.

Since

$$\frac{A_A}{A_{AP}} = \frac{A_B}{A_{AP}} = \frac{1}{2}$$

and if $F_s/A_{AP}$ is called the spring pressure it can be said that

if $(P_{AP} + P_s) > \frac{1}{2}(P_A + P_B)$     the poppet is closed,

and

if $(P_{AP} + P_s) < \frac{1}{2}(P_A + P_B)$     the poppet is open.

Flow forces across the poppet are generally small and can be neglected at this stage.

### 3.1.2 Springs Available
Three springs are available for the 2:1 poppet, giving spring pressures as follows:

| Spring | Spring pressure, $P_s$ (bar) |
|--------|------------------------------|
| Light  | 0·25 |
| Medium | 1·25 |
| Heavy  | 2·50 |

### 3.1.3 Cracking Pressure
In the context of cartridge valves, it is confusing to talk in terms of 'cracking pressure'. With a conventional check valve, spring pressure and cracking pressure mean the same thing, i.e. the pressure required at the inlet over and above the pressure at the outlet when the valve is just about to open. With a cartridge valve both the inlet and outlet pressure will tend to open the poppet. In the case of a 2:1 poppet, 'cracking pressure' is in fact twice the spring pressure, but this will not be true of other-ratio poppets. Spring pressure is therefore defined as 'the pressure that would be required on the Ap port of the cartridge to exert the same force as the spring'.

### 3.1.4 Opening Pressure Calculation
Determine what pressure is required on the A port of the cartridge shown in Fig. 6 in order to open it.

$$\text{Pressure opening poppet} = \tfrac{1}{2}(P_A + P_B)$$
$$= \tfrac{1}{2}P_A + \tfrac{1}{2}P_B$$
$$= \tfrac{1}{2}P_A + 2\cdot5 \qquad \text{(i.e. } \tfrac{1}{2} \times 5)$$

$$\text{Pressure closing poppet} = P_{AP} + P_s$$
$$= 20 + 2\cdot5 \qquad \text{(2·5 bar spring pressure)}$$
$$= 22\cdot5 \text{ bar}$$

At the time when the poppet is just about to open, opening pressure = closing pressure, i.e.

$$\tfrac{1}{2}P_A + 2\cdot5 = 22\cdot5$$

so

$$P_A = 40 \text{ bar}$$

## 3.2 The 1·1:1 Ratio Poppet
In this version the diameter of the seat is increased relative to the diameter

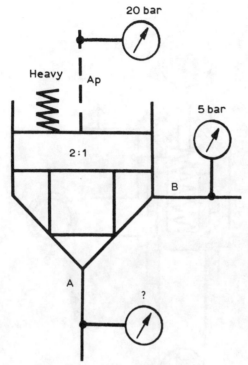

FIG. 6. 2:1 ratio poppet example.

of the poppet. This has the effect of increasing the A port area and decreasing the B port area, as shown in Fig. 7.

*3.2.1 Ratio of Areas*

The ratio of areas is now approximately

$$A_{AP}:A_A:A_B = 1\cdot1:1:0\cdot1$$

In most cases this can be approximated to

$$A_{AP}:A_A:A_B = 1:0\cdot9:0\cdot1$$

Considering the opening and closing forces on the poppet as before, it can be shown that

if $(P_{AP} + P_s) > (0\cdot9P_A + 0\cdot1P_B)$ the poppet is closed,
and
if $(P_{AP} + P_s) < (0\cdot9P_A + 0\cdot1P_B)$ the poppet is open

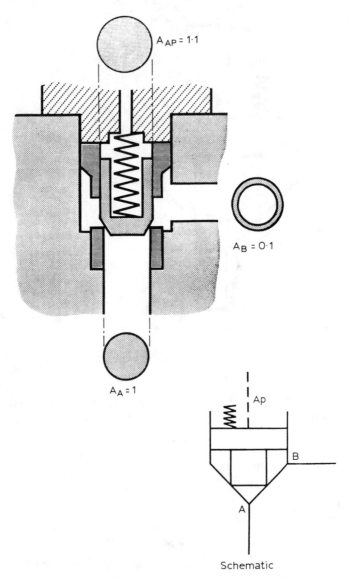

FIG. 7.   The 1·1:1 ratio poppet.

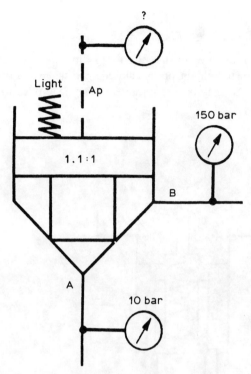

FIG. 8. 1·1:1 ratio poppet example.

The same three springs are used in the 1·1:1 poppet, i.e. spring pressures of 0·25, 1·25 and 2·5 bar.

### 3.2.2 Closing Pressure Calculations

Determine the minimum pressure required on the Ap port of the cartridge in Fig. 8 to hold the poppet closed.

$$\text{Pressure opening poppet} = 0.9P_A + 0.1P_B$$
$$= (0.9 \times 10) + (0.1 \times 150)$$
$$= 24$$

$$\text{Pressure closing poppet} = P_{AP} + P_s$$
$$= P_{AP} + 0.25$$

At the cracking point

$$P_{AP} + 0.25 = 24$$

so

$$P_{AP(min.)} = 23 \cdot 75 \text{ bar}$$

As the above illustrates, the use of a 1·1:1 ratio poppet allows a relatively low pilot pressure to hold the poppet closed against a higher pressure acting on the B port. This may be a requirement in some applications, as will be shown later.

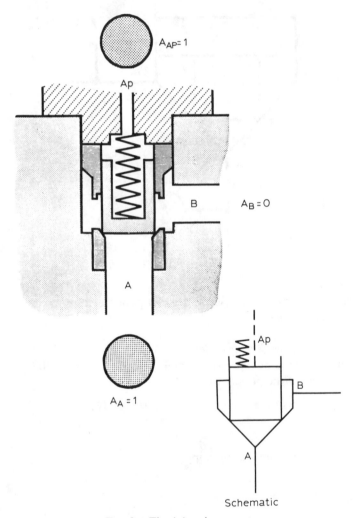

Schematic

FIG. 9.   The 1:1 ratio poppet.

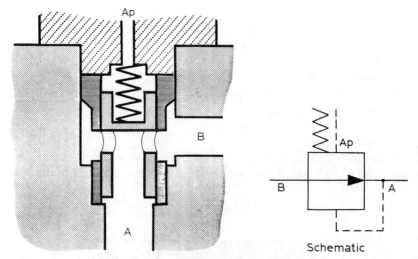

FIG. 10.   The 1:1 ratio poppet normally open.

### 3.3 The 1:1 Ratio Poppets

The poppets used for pressure control valve functions have a 1:1 ratio, i.e. the area of the A port is exactly equal to the area of the Ap port, which means that the B port area is zero (Fig. 9). Since the B port has no active area, it follows that

$$\text{if } (P_{AP} + P_s) < P_A \qquad \text{the poppet is open,}$$
and
$$\text{if } (P_{AP} + P_s) > P_A \qquad \text{the poppet is closed.}$$

A second type of 1:1 poppet is the normally open version shown in Fig. 10, and is used for pressure-reducing valve applications. It can also be used as a hydrostat, i.e. to maintain a constant pressure difference across a throttle valve.

Modified versions of the poppets discussed are used in certain applications and these will be described in Section 4.

## 4  VALVE FUNCTIONS

The range of control valve functions is achieved basically by controlling the pressure on the Ap port of the appropriate poppet. Pilot valves may be either incorporated in, or mounted on, the cover plate. In order to control the opening and closing speed of the poppet, orifice plugs can be fitted into the pilot connections. The orifice plugs are readily accessible, which enables the system to be tuned during commissioning.

### 4.1 Check Valve Functions
Using a basic cover and a 2:1 poppet provides a check (non-return) valve function as shown in Fig. 11.

#### 4.1.1 Basic Arrangement
The Ap port of the poppet is connected by drillings in the cover plate and manifold block to the B port, with a damping orifice plug incorporated in

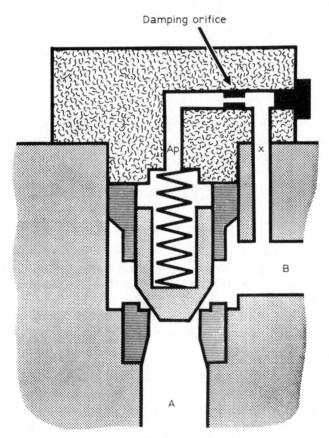

FIG. 11.   Check valve function.

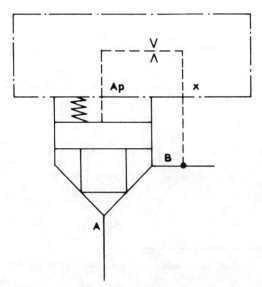

FIG. 12. Check valve function schematic.

the cover plate. This arrangement would be represented schematically as shown in Fig. 12.

Assuming that the poppet ratio is 2:1 then, as shown before, for the poppet to be open

i.e.

$\frac{1}{2}(P_A + P_B)$ must be greater than $P_s + P_B$ (i.e. $P_{AP} = P_B$)

$$P_A > (2P_s + P_B)$$

If $P_B > P_A$ then the poppet cannot open, so flow through the valve is only possible from A to B. Using a conventional check valve, the equivalent arrangement would be as shown in Fig. 13.

A ⟨○⟩www B    FIG. 13. Check valve function
2P_s                    schematic.

### 4.1.2 Alternative Arrangement

It would also be possible to achieve a check valve function by connecting the Ap port to the A port as in Fig. 14. This would then give free flow from B to A and blocked flow from A to B.

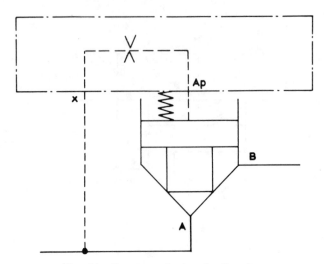

FIG. 14.   Incorrect check valve function.

It must be remembered, however, that leakage can occur between the Ap port and the B port (refer to Fig. 4). In the reverse (blocked) flow direction, therefore, leakage may exist between the A and B ports (via the Ap port). To avoid this, the configuration shown in Fig. 12 should always be used for check valve functions.

### 4.1.3 Solenoid-operated Check Valve

Incorporating a solenoid directional valve into the pilot line of Fig. 12 gives effectively a 'solenoid-operated' check valve (Fig. 15).

With the solenoid de-energised Ap is connected to B, which gives the check valve function as in Fig. 12, i.e. free flow A to B, blocked flow B to A. Energising the solenoid connects the Ap port to tank. In this situation only sufficient pressure to overcome the spring is required in either the A or B ports to open the poppet. With the solenoid energised, therefore, the valve gives virtually free flow A to B or B to A.

If the solenoid valve is a sliding spool valve, then leakage across this valve may have to be considered. If the solenoid valve is de-energised, flow in the reverse direction B to A is blocked and a virtually leakproof seal exists between B and A. Leakage will occur, however, from the B port across the spool of the solenoid valve and back to tank. This fact may have to be taken into account in the design of the system.

Provided the pilot flow rates are not too great, a poppet-type solenoid

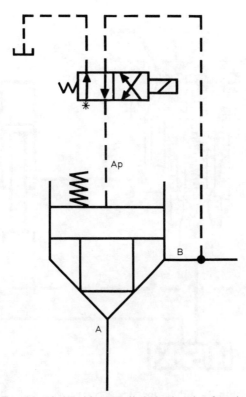

FIG. 15. Solenoid controlled check valve function.

pilot valve can be used to overcome this problem. The poppet solenoid valve is a positive seating (and hence leakproof) valve, thus preventing leakage from the B port to tank.

### 4.1.4 Pilot-operated Check Valve

The cartridge version of a pilot-operated check valve uses a 2:1 poppet and a special cover plate as illustrated in Fig. 16. The use of pilot-operated (PO) check valves will not be as widespread in cartridge, as opposed to conventional, valve systems, since the main directional valve itself may be a poppet-type (positive seating) valve. There are times, however, when a PO check valve is useful, for example to avoid pilot valve leakage, or where a spool-type main directional valve is used.

Figure 16 shows a cartridge PO check valve in conjunction with a spool-type directional valve.

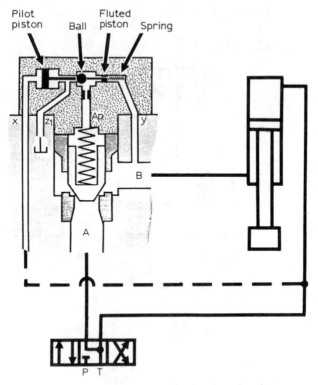

FIG. 16.    Pilot-operated check valve, closed.

*4.1.4.1 Centre position.* With the directional valve in the centre position as drawn, the left-hand side of the pilot piston is open to tank via the directional valve. Load pressure from the B port of the cartridge, together with the spring on the fluted piston, push the ball on to the left-hand seat of the cover plate. This allows load pressure to be connected to the top of the cartridge poppet, holding it on to its seat, and the valve is closed. The ball and seat arrangement in the cover plate prevents any leakage through the pilot stage.

*4.1.4.2 Piloted open position.* Figure 17 shows the valve in the piloted open position. The spool valve has moved across to connect pressure to the top of the cylinder and also to the left-hand side of the pilot piston in the cover plate. The pilot piston therefore moves to the right and pushes the ball on to the right-hand seat. The top of the cartridge poppet is now open to tank via the internal connections in the cover plate and so, provided the

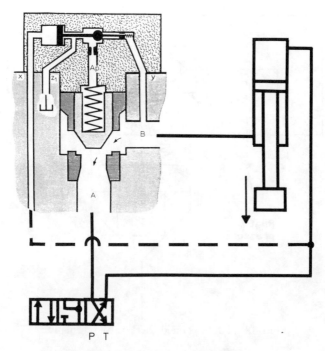

FIG. 17. Pilot-operated check valve, piloted open.

pressure at the B port is high enough to overcome the spring, the poppet will open, allowing the cylinder to descend. Flow up the pilot connection from the B port is blocked by the ball being pushed on to the right-hand seat.

In practice, as with a conventional valve system, the cylinder would have to be prevented from running away with itself on the down stroke, by using a flow control or counterbalance valve, for example.

The pilot pressure to open the valve must be at least 30% of the load pressure, including any intensification. Figure 18 shows the schematic representation of a pilot-operated check valve.

## 4.2 Directional Valve Functions

### 4.2.1 Basic Construction

Cartridge valves used for directional functions generally incorporate solenoid pilot valves to open or close the poppet as required. The solenoid valve may be mounted on top of the cover plate, as shown in Fig. 19. For cartridge sizes up to and including NG40 a CETOP 3 (CETOP is the

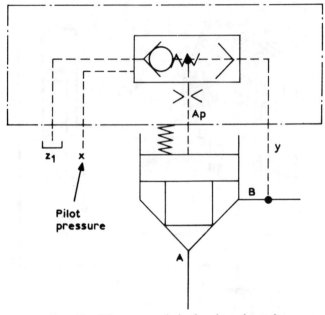

FIG. 18.    Pilot-operated check valve schematic.

standard organisation for fluid power) size solenoid pilot is used, and CETOP 5 for the larger sizes.

Figure 19 also shows the orifice plugs which may be fitted into the cover plate in order to control the opening or closing speed of the poppet. A system can therefore be tuned during commissioning (to eliminate shocks, etc.) by changing the size of the orifice plugs which are easily accessible from the sides of the cover plate.

### 4.2.2  Pilot Connections
Figure 20 illustrates the schematic arrangement for a cartridge insert and CETOP 3 size cover plate and solenoid valve. This is used for cartridge sizes from NG 16 to NG 40. The bottom surface of the cover plate has five pilot connections, as follows:

(1)   Ap port—connects to top of cartridge poppet;
(2)   x port—connects to P port of solenoid valve;
(3)   y port—connects to T port of solenoid valve;
(4)   $z_1$ port—connects to A port of solenoid valve;
(5)   $z_2$ port—connects to B port of solenoid valve.

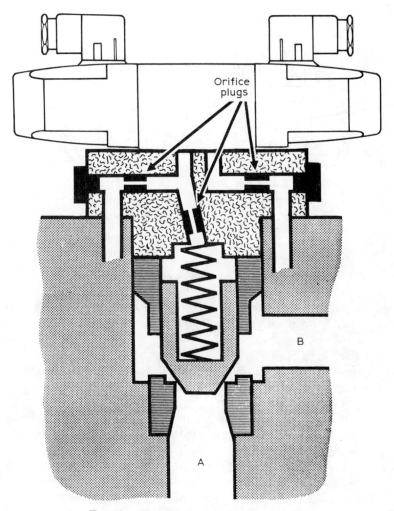

FIG. 19. Cartridge insert with solenoid pilot.

A standard orifice is fitted into the cover plate, in this case in the $z_1$ port. The diameter of the orifice fitted will vary depending on the size of insert used. Additional tapped ports are provided in the Ap, x, $z_1$ and $z_2$ ports for fitting additional orifice plugs if required.

It is possible, of course, to fit any type of solenoid valve arrangement on the cover plate, i.e. spring offset, detented or spring centered, depending

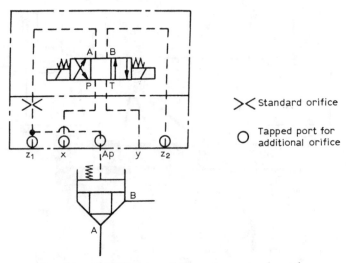

Standard orifice

Tapped port for
additional orifice

FIG. 20.    CETOP 3 size pilot arrangement: schematic.

upon the circuitry requirements. Lever- or cam-operated pilot valves may
also be used for manual/mechanical control of the cartridge.

### 4.2.3 Examples of Use
An example of the use of solenoid-controlled cartridges for directional
valve functions is shown in Figs 21(a) and 21(b).

#### 4.2.3.1 Operational details. The cylinder is controlled by means of four
cartridges, of which one incorporates the solenoid pilot valve on its cover
plate and the other three use a standard cover.

The circuit is shown simplified in Fig. 21(a) and drawn in full in Fig. 21(b).

With the solenoid valve in the centre position, and assuming pressure is
maintained in the P line, then system pressure is applied to the top of each
cartridge poppet. This pressure, together with the spring, pushes each
poppet on to its seat and all cylinder lines are blocked.

Energising solenoid S1 now connects system pressure to the top of
poppets 2 and 4, holding them closed, and connects the top of poppets 1 and
3 to tank. Only a small amount of pressure to overcome the spring is
required at the B port of cartridge 3 and the A port of cartridge 1 in order to
open them. Flow can now take place from the P line via cartridge 3 into the
annulus side of the cylinder, and exhaust flow from the full-bore side of the
cylinder via cartridge 1 to tank. The cylinder therefore retracts.

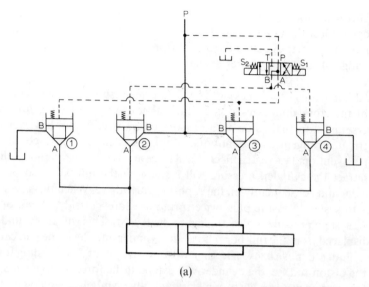

(a)

FIG. 21(a). 'Four-way' valve, simplified circuit.

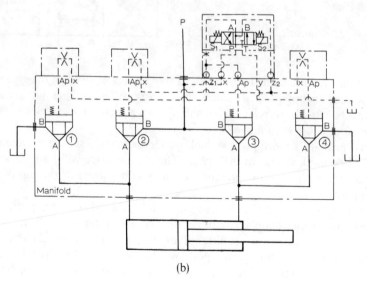

(b)

FIG. 21(b). 'Four-way' valve, full circuit.

Energising solenoid S2 holds cartridges 1 and 3 closed and allows 2 and 4 to open. Flow then takes place from the P line to the full-bore side of the cylinder (via cartridge 2) and exhaust from the annulus side to tank (via cartridge 4), and the cylinder extends.

### 4.2.3.2 More positive locking.

The arrangement shown in Figs 21(a) and 21(b) therefore duplicates a conventional four-way valve, i.e. forward, reverse and stop. In the stop position (solenoid valve centered), however, the cartridge arrangement does not quite duplicate a 'closed centre' condition of a conventional valve. If a mechanical force acts on the piston rod tending to retract the cylinder, pressure will increase in the full-bore end of the cylinder and consequently on the A port of cartridge 2. When this pressure rises to system pressure plus twice spring pressure, cartridge 2 will open (since system pressure is also applied to the B port). This will allow fluid to be displaced from the full-bore end of the cylinder into the P line and cause the cylinder to move back. The same will also be true for an extending force on the piston rod, i.e. the cylinder will extend if the force is large enough.

If the system were drawn in conventional valve symbols, it would give the arrangement shown in Fig. 22.

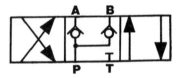

FIG. 22.   Equivalent   'conventional'
symbol.

If the pressure in the P line drops to zero, for example when the system is unloaded or shut down, then only spring force will be holding the cartridge poppets closed. A relatively small force applied to the cylinder will therefore cause it to move.

To provide a more positive lock on the cylinder a shuttle valve cover could be used, as shown in Fig. 23. The shuttle valve will automatically direct the highest of the pressures in the x and y ports to the top of the poppet.

Referring again to Fig. 21(b), suppose that in the cylinder stop position a mechanical force is applied to the piston rod which is tending to retract it. If cartridge 2 is fitted with a shuttle valve cover, as shown in Fig. 24, then the highest pressure in the x and y ports will be used to hold poppet 2 closed. The y port is connected to system pressure and the x port to the cylinder full-bore pressure, so if the force on the cylinder creates a pressure higher

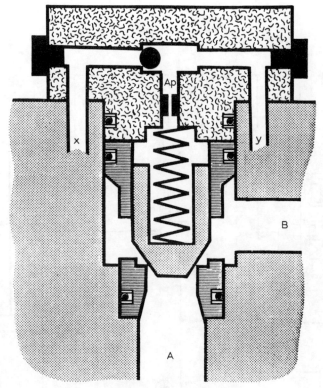

FIG. 23. Shuttle valve cover.

than system pressure, then the full-bore pressure is used to hold cartridge 2 closed and thus prevent cylinder movement. If the mechanical force is large enough such that the pressure created in the full bore of the cylinder exceeds twice system pressure, then cartridge 1 may open again, allowing the cylinder to move. In certain cases this may be desirable, i.e. to relieve shock loads on the cylinder, but if not it can be overcome by fitting cartridge 1 with a shuttle valve cover or by using a 'blocked centre' circuit, as described later.

*4.2.3.3 Different flow configurations.* Shuttle valve covers are also available with a CETOP 3 interface, as shown in Fig. 25. It must be remembered, however, that because the shuttle valve uses two of the pilot connections in the cover plate (x and y) it is only possible to use three of the four ports of the solenoid valve, i.e. the B port cannot be used.

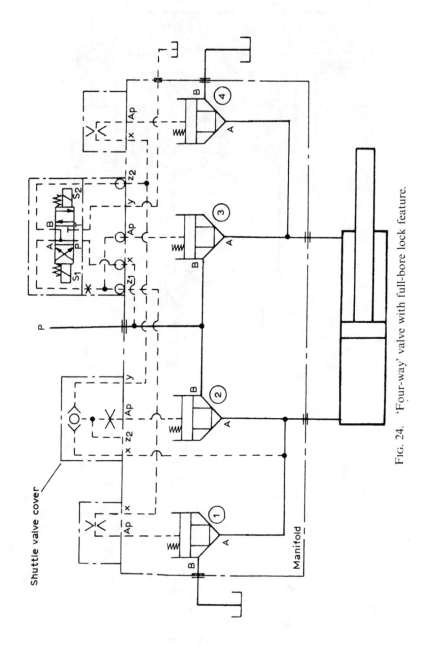

Fig. 24. 'Four-way' valve with full-bore lock feature.

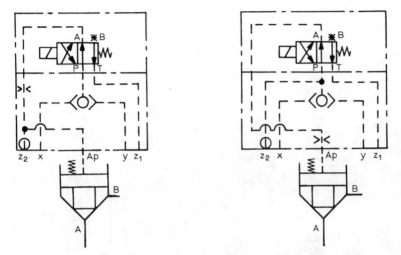

FIG. 25. Shuttle valve covers with CETOP 3 interface.

By controlling each cartridge with an independent solenoid pilot valve (Fig. 26) it is possible to get 16 arrangements of solenoid energisation, giving 12 different flow configurations, i.e. equivalent to a 12-position sliding spool valve. Also each cartridge can be controlled independently of the others to achieve different transient conditions; for example, one side of the cylinder can be decompressed before pressure is applied to the other side.

A further advantage of the 'four-way' cartridge arrangement is that each cartridge can be sized to its own particular flow rate. For example, supposing the pump flow in Fig. 26 is 100 litre min$^{-1}$ and the cylinder piston ratio is 2:1, cartridges 2 and 3 would be sized for 100 litre min$^{-1}$, cartridge 1 for 200 litre min$^{-1}$ and cartridge 4 for 50 litre min$^{-1}$ (assuming regeneration is not used).

## 4.3 Relief Valve Functions

The 1:1 poppet shown in Fig. 9 is used for relief valve functions, and the basic relief valve configuration is shown in Fig. 27.

### 4.3.1 Relief Valve Operation

The operation of the cartridge relief valve is very similar to that of a conventional two-stage relief valve. The A port is the pressure inlet and the B port the tank return. A pilot relief valve is incorporated in the cover plate

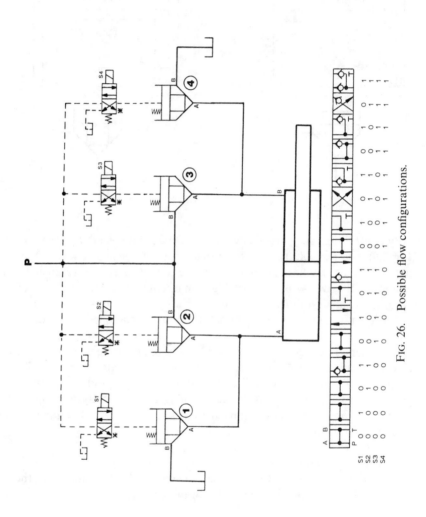

FIG. 26.   Possible flow configurations.

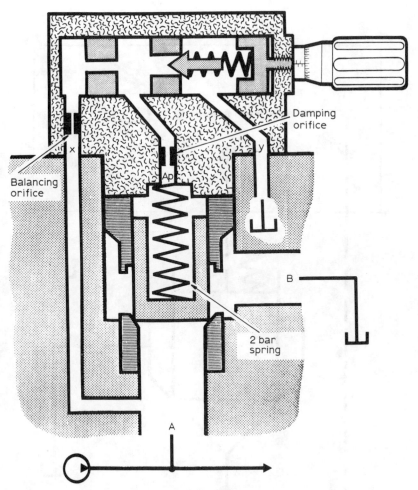

FIG. 27.  Relief valve arrangement.

and is fed via a balancing orifice from the inlet port. (The balancing orifice corresponds to the hole through the hydrocone piston of a conventional relief valve.)

When the A port pressure (system pressure) is less than the setting of the pilot relief, no flow exists across the balancing orifice. This means that the pressure at the top and bottom of the cartridge poppet is the same so the spring holds it on to its seat and the valve is closed.

When system pressure reaches the setting of the pilot relief, the small

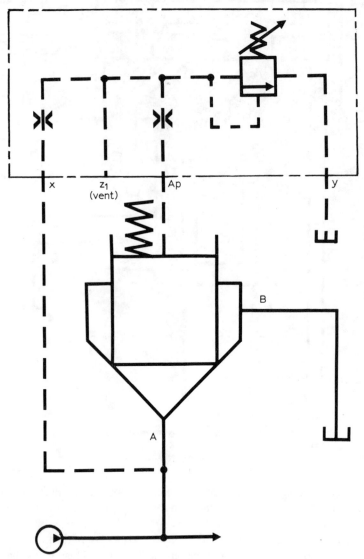

FIG. 28.    Relief valve: schematic.

conical poppet is lifted off its seat and flow passes across it to tank via port y. This flow creates a pressure difference across the balancing orifice and therefore also across the cartridge poppet. A pressure difference of 2 bar will be sufficient to overcome the spring in the poppet, causing it to lift and the valve to open.

The opening (and closing) speed of the poppet is determined by the orifice in the Ap port. The pilot drain port, y, may be connected to the main tank return port, B, or drained separately to tank to avoid the effect of any back pressure in the main tank return line.

Also included in the cover plate is a vent port (port $z_1$ in Fig. 28), which acts in an identical way to that on a conventional relief valve. With the vent port blocked, normal relief valve operation occurs. With the vent port open to tank, the relief valve will open at approximately 2 bar (vented).

### 4.3.2 Increasing the Range

Again, as with conventional valves, cover plates are available for mounting additional pilot valves on top to provide a range of relief valve controls. Figure 29(a) illustrates a solenoid venting feature. With the solenoid valve de-energised as drawn, the top of the cartridge poppet (effectively the vent port) is connected to tank via the solenoid valve and port y. The relief valve will open therefore at approximately 2 bar on the inlet port and the system is unloaded. Energising the solenoid valve blocks the vent port and the relief valve reverts back to the full pressure setting of the pilot relief valve.

Using a left-hand solenoid valve will give the opposite effect, i.e. energise to off-load, de-energise to load.

### 4.3.3 Two-pressure Arrangement

Figure 29(b) shows a two-pressure arrangement. With the left-hand solenoid energised, the pressure on top of the poppet (and hence system pressure) is controlled by pilot relief valve 1. Energising the right-hand solenoid means that system pressure is now controlled by pilot relief valve 2 (provided it is set lower than 1). If both solenoids are de-energised, the top of the poppet is connected to tank and the relief valve is unloaded.

By using an electrically modulated pilot relief valve mounted on the cover plate (Fig. 30) it is possible to achieve remote electronic control of the relief valve cartridge.

Varying the electrical signal to the modulating relief valve adjusts its pressure setting and hence the pressure setting of the main relief valve cartridge. Maximum pressure is determined by the setting of the manually adjusted pilot relief valve.

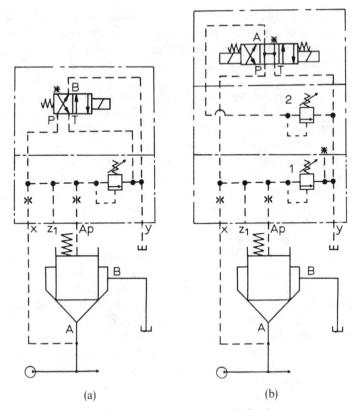

(a)                                    (b)

FIG. 29.   Solenoid-controlled relief valves.

A second version of the modulating relief valve may also be used which incorporates an armature position feedback device. Using a closed-loop control signal improves the hysteresis and repeatability of the valve for high-performance applications.

When using either of the electrically modulated versions of the valve, the drain port (y) should be connected directly to tank to avoid the possibility of any back pressure in this line.

## 4.4  Reducing Valve Functions
The pressure-reducing valve cartridge uses a sliding (as opposed to seated) poppet which is normally open. The B port is the high-pressure inlet and the A port the reduced-pressure outlet.

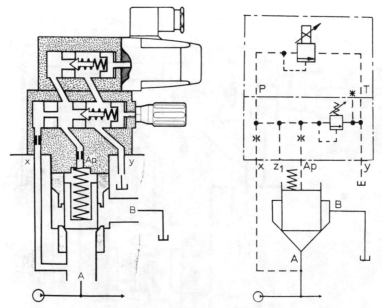

FIG. 30. Electrically modulated relief valve.

### 4.4.1 Pilot Stage

Referring to Fig. 31, the pilot relief valve in the cover plate is fed with oil from the high-pressure inlet port B via port x. In any application of the valve the inlet pressure may vary, which means that a fixed orifice in the x port would produce a varying flow to the pilot relief poppet and hence cause problems of pressure override in the pilot stage. To avoid this, what is in effect a fixed pressure-compensated flow control valve is incorporated in the x port. This consists of a spring-loaded sliding piston with a fixed orifice in one end. The other end of the piston produces a variable orifice as it slides in its sleeve.

The pressure in the Ap port (i.e. the pressure on the top of the main cartridge poppet) is determined by the setting of the pilot relief valve. The object of the pressure-compensated flow control is to maintain a constant flow to the pilot relief valve, irrespective of inlet pressure, and hence maintain a constant pressure in the Ap port. If a rise in inlet pressure creates a higher pilot flow, the pressure drop across the fixed orifice in the sliding piston increases. This will push the piston to the left, which tends to close the variable orifice and thus reduce the pilot flow to its original value. Irrespective of the valve setting or the inlet port pressure, therefore, the

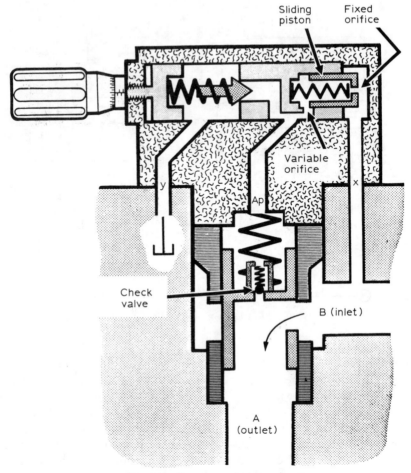

FIG. 31.   Pressure-reducing valve.

pilot flow remains constant, thereby avoiding pressure override problems
with the pilot relief valve.

### 4.4.2  Pressure Overshoot
As main flow passes through the valve from B to A (Fig. 31) the cartridge
poppet will start to close off when the outlet port pressure equals the Ap port
pressure plus the main poppet spring pressure. The poppet will therefore
close sufficiently to reduce the inlet pressure to this value.

   To take account of any pressure overshoots that may occur (for example
when a cylinder reaches the end of its travel) a small check valve is

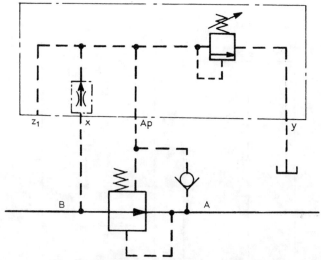

FIG. 32. Pressure-reducing valve: schematic.

incorporated in the main poppet. Any overshoot in outlet port pressure will open this check valve and relieve via the pilot stage.

As with most pressure-reducing valves, the drain connection y should be connected directly and separately to tank.

Figure 32 illustrates the schematic arrangement of the cartridge reducing valve. As with the pressure relief valve, the $z_1$ port may be used for remote control functions. Figure 33(a) shows a two-pressure option (achieved by switching from one pilot relief valve to the other), and Fig. 33(b) shows the electrically modulated version.

## 4.5 Flow Control Functions

By using a restrictor-type poppet, i.e. a standard 2:1 poppet with a notched skirt, and a cover plate fitted with a stroke adjuster a throttling feature can be added to the cartridge (Fig. 34). In this case flow restriction will occur in both directions, i.e. A to B and B to A. This arrangement therefore gives a simple needle valve or non-compensated type of flow restrictor.

### 4.5.1 A Four-way Application

Applying the restrictor poppet to a 'four-way' circuit as shown in Fig. 35.

Restrictor poppet 1 gives meter-out control in the cylinder retract direction.

Restrictor poppet 2 gives meter-in control in the extend direction.

Restrictor poppet 3 gives meter-in control in the retract direction.

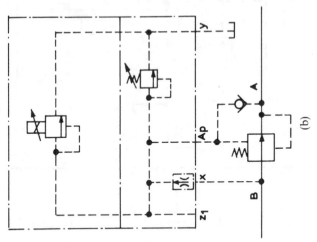

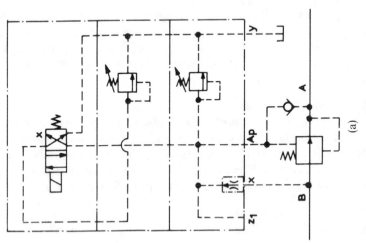

FIG. 33.   Two-pressure (a) and electrically modulated (b) reducers.

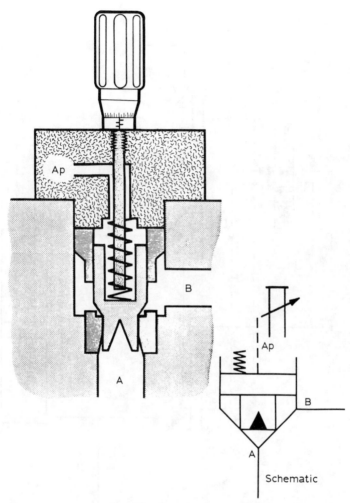

FIG. 34. Restrictor poppet.

If a restrictor poppet is used in position 4 to provide meter-out control in the extend direction, then a problem may occur. Because of the differential area across the cylinder piston, throttling the annulus flow from the cylinder when it is extending creates an intensified pressure in this part of the system. This intensified pressure will act on the A port of cartridge 3, and since the B port is also subjected to system pressure cartridge 3 may open. This will create a partial regenerative circuit, i.e. cause the cylinder

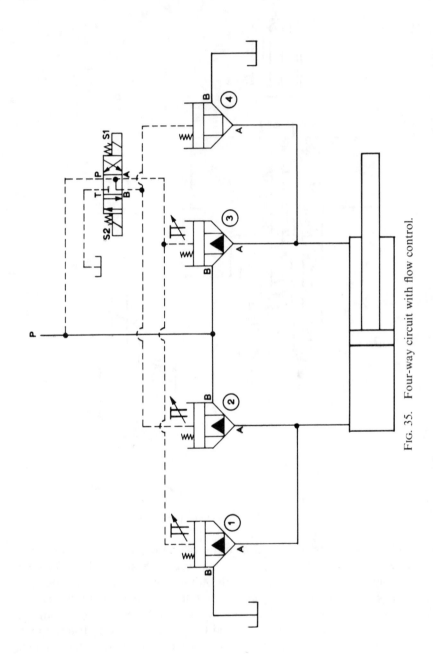

FIG. 35.    Four-way circuit with flow control.

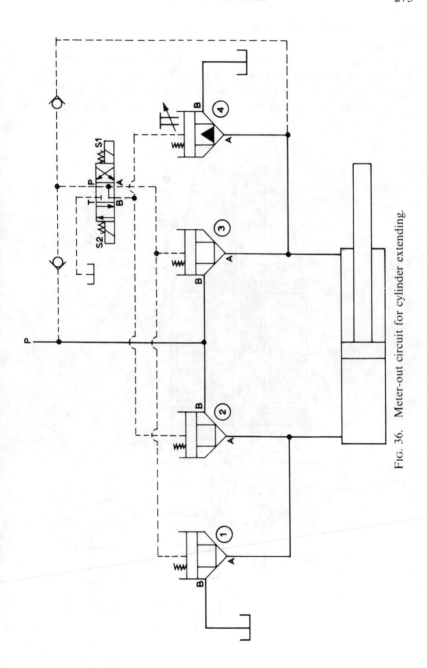

Fig. 36. Meter-out circuit for cylinder extending.

speed to increase. The more cartridge 4 is closed off, the faster the cylinder will move.

One method of overcoming this is to use the modified circuit shown in Fig. 36 (a second alternative is given later). By using two check valves (or one shuttle valve) in the pilot feed to the cartridges, if the intensified pressure in the annulus end of the cylinder exceeds system pressure, then the

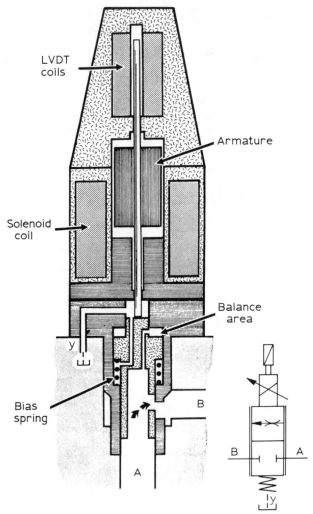

FIG. 37.  Electrically modulated throttle, direct-acting.

intensified pressure is used to hold cartridge 3 closed. This ensures that cartridge 3 cannot open during the extend stroke of the cylinder but, as with any meter-out circuit, the amount of pressure intensification should be checked to make sure that it cannot reach dangerous proportions.

### 4.5.2  Electrically Modulated Throttle Valve

Figure 37 shows a direct-acting electrically modulated throttle valve. Flow through the valve is in this case in the A to B direction only and restriction is created by varying the position of the spool relative to the sleeve. It should be noted therefore that this is not now a positive seating valve and leakage may have to be taken into consideration. The spool is positioned in the sleeve by means of a proportional solenoid and an in-built LVDT (linear variable differential transformer) provides a spool position feedback signal.

The spool is hydraulically balanced by means of a drilling which connects the inlet (A port) area to an annulus-shaped area (equal to the A port area) on the top of the spool. A spring biases the spool to the closed position and so the solenoid force only has to overcome the spring force.

The solenoid is used with a closed-loop electronic control, whereby an input signal continuously opens the valve until the required amount of opening is achieved, i.e. when the LVDT feedback signal coincides with the input signal. The valve opening is therefore proportional to the solenoid input signal. In its basic form the valve is non-pressure-compensated, but may be incorporated with a hydrostat or flow sensor to provide pressure compensation, as will be described later.

The direct-acting throttle is used for the NG 16 size of cartridge. Larger cartridge sizes use a two-stage arrangement, as shown in Fig. 38.

In this case the proportional solenoid acts directly on to a spring-loaded pilot spool. A pilot flow is taken from the inlet (A) port of the valve and is bled to tank via the port that connects to the top of the main poppet. With zero signal to the solenoid, the spring pushes the pilot spool to the fully-up position. This blocks the pilot bleed flow and so full inlet port pressure is applied to the top of the main poppet. With the same pressure at the top and bottom of the main poppet, its spring pushes it to the fully-closed position.

As a signal is applied to the solenoid the pilot spool moves down. This then opens up a bleed path for the pilot flow and thereby reduces the pressure on the top of the main poppet. This pressure drop causes the poppet to lift, i.e. opening the main flow path through the valve until the feedback signal from the position sensor coincides with the input signal to the solenoid. The valve then reaches a steady-state position at an opening determined by the input signal.

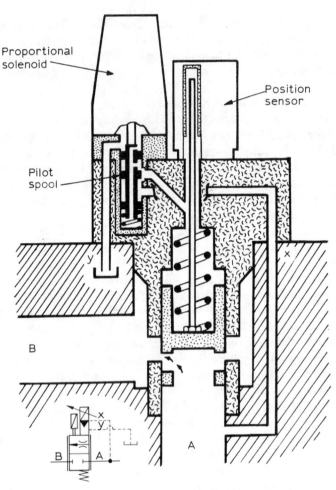

FIG. 38.    Two-stage electrically modulated throttle.

The pilot supply for the valve must be taken directly from the A port, and be at a minimum pressure of 10 bar.

As with the direct-acting version, the valve is non-pressure-compensated and must be used in conjunction with either a hydrostat or a flow sensor if pressure compensation is required.

### 4.5.3  Hydrostats

By using a hydrostat with any of the flow control valves mentioned either a

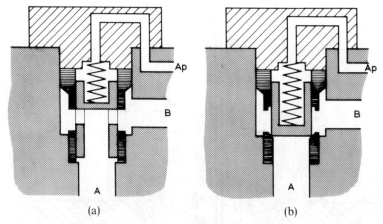

(a)                                    (b)

FIG. 39.    (a) Spool and (b) poppet-type hydrostats.

two- or three-port pressure-compensated flow control function can be achieved. Two-port valve arrangements use a spool-type (normally open) hydrostat, which can provide either a fixed pressure differential [Fig. 39(a)] or an adjustable differential if a complete reducing valve assembly is used. The circuitry arrangements are shown in Fig. 41.

Three-port or spill-off-type flow controls use a poppet-type hydrostat. The fixed differential version is shown in Fig. 39(b) and a variable differential is provided by using a complete relief valve assembly (see Fig. 41).

### 4.5.4 Flow Sensor

An alternative method of obtaining pressure compensation is to use an electronic flow sensor in conjunction with a proportional throttle. The flow sensor provides a feedback signal proportional to the controlled flow which can be used to modulate the input signal to the throttle, thus compensating for variations in load or pressure.

Figure 40 shows the flow sensor construction and Fig. 41 illustrates its applications.

As flow passes through the sensor (in the A to B direction) the spool is lifted against the spring an amount corresponding to the flow rate. The spool position sensor then provides a feedback signal proportional to spool lift (and hence flow rate) to the electronic control unit.

One advantage with using a flow sensor is that it need not be mounted in-line with the flow control valve itself. For example, if a bleed-off type of flow

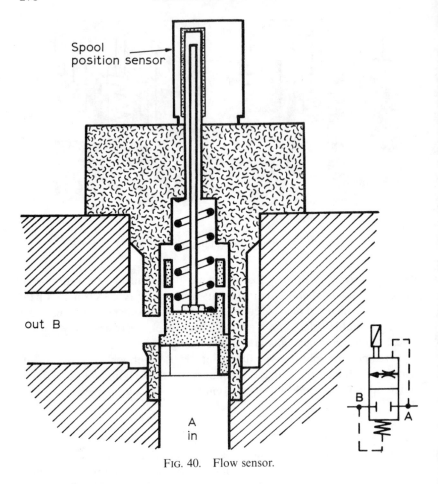

FIG. 40.    Flow sensor.

control is used, the flow sensor can be mounted in the 'meter-in' line, thus compensating for slight pump flow variations.

### 4.5.5 Flow Control Applications

Figure 41 illustrates some of the possible combinations of the components described.

(a)  This arrangement uses a restrictor poppet (1) with a spool-type (normally open) hydrostat (2). The spring on the hydrostat spool maintains a constant 10-bar pressure difference across the restrictor poppet, thus generating a pressure-compensated flow control.

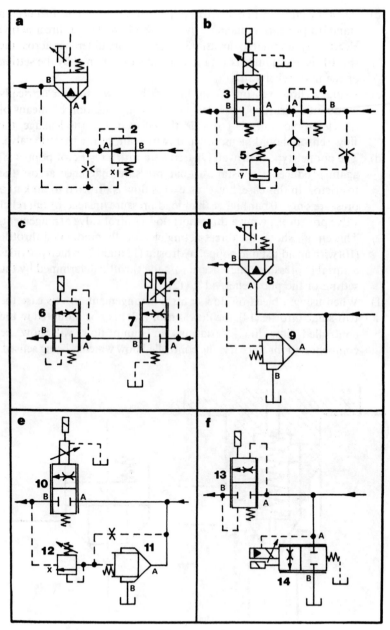

FIG. 41. Flow control applications.

(b)  An adjustable in-line hydrostat (4 and 5), which is in effect a standard pressure reducing valve, is used here with a direct-acting electrically modulated throttle (3). The pressure difference across the throttle is maintained constant at a value determined by the setting of the pilot relief valve (5).

(c)  Using a flow sensor (6) in line with a two-stage electrically modulated throttle (7) creates pressure compensation by means of the electronic control. To avoid the effect of throttle leakage, the flow sensor should be mounted downstream of the throttle valve.

(d)  The poppet-type hydrostat (9) used here with a restrictor poppet (8) again produces a constant 10-bar pressure difference across the restrictor. In this case, however, excess flow is diverted to tank at a pressure only 10 bar higher than load pressure (instead of full relief valve pressure), giving a three-port flow control valve arrangement.

(e)  This circuit shows the direct-acting electrically modulated throttle (10) with an adjustable poppet hydrostat (11 and 12), which provides a variable pressure difference across the throttle determined by the setting of the pilot relief valve (12).

(f)  When using a bleed-off flow control arrangement, in this case the two-stage throttle (14), the flow sensor (13) may be mounted in the controlled flow line. Variations in pump flow will now be compensated for since it is the controlled flow which is being sensed.

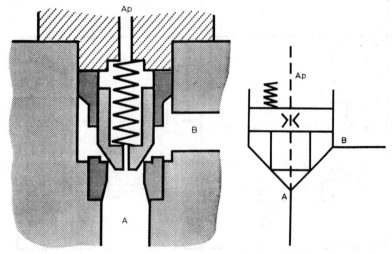

FIG. 42.  Orifice poppet.

### 4.6 Miscellaneous Functions

An option on the 1·1:1 poppet is an orifice plug fitted into the poppet itself (Fig. 42). This allows pilot flow through the poppet and can be used in safety or fast sequencing circuits.

Also used in safety circuits is the position-monitoring option shown in Fig. 43. A proximity switch fitted to the lower cover plate monitors the open or closed position of the valve. The switch is operated by means of a plunger which is pressure-biased (by means of area difference) against the cartridge poppet. (Smaller valve sizes use a spring-biased plunger.)

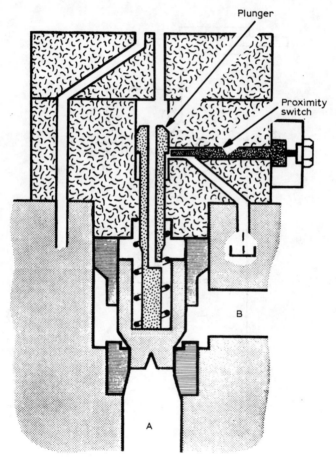

FIG. 43. Position-monitoring option.

The pilot valve fitted to the top cover may also be fitted with a spool position switch so, depending upon requirements, either one or two levels of position monitoring can be provided.

## 5 CIRCUITS

This section covers some of the many different ways in which cartridge valves can be used to achieve different circuitry arrangements. If a designer is more familiar with conventional sliding spool valves, there is a tendency

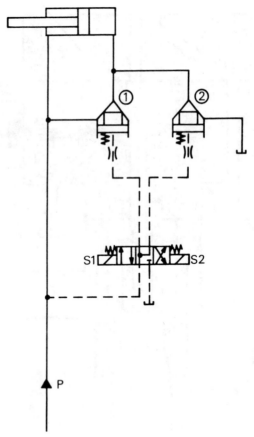

FIG. 44.   Regenerative circuit.

when designing cartridge systems to think it out in conventional valves and then convert to cartridge. This should be avoided, however, as it tends to result in over-complicated systems. Very often a cartridge system requires fewer valves to achieve the same functions than a conventional system.

### 5.1 Regenerative Circuit

With both solenoids de-energised (Fig. 44) system pressure is applied to the pilot ports of both cartridges, holding them closed. The piston will not move, therefore, unless a force is exerted on the piston rod large enough to open cartridge 1. (Cartridge 1 already has system pressure acting on its B port, so if a force on the cylinder creates a pressure equal to system pressure plus spring pressure in the full-bore side, cartridge 1 will open, allowing the cylinder to retract.)

Energising solenoid S1 holds cartridge 1 closed and allows cartridge 2 to open. Pump flow then enters the annulus side of the cylinder and exhaust flow from the full-bore side is directed to tank via cartridge 2, and the piston retracts.

Energising solenoid S2 holds cartridge 2 closed and allows cartridge 1 to open. The differential areas across the piston now cause it to extend with the exhaust flow from the annulus side adding to the pump flow.

### 5.2 Regenerative Circuit with Full Force at End of Stroke

With both solenoids de-energised (Fig. 45) pressure is applied to the pilot connections of all cartridges, holding them closed and preventing cylinder movement (again assuming no large external forces).

Energising S1 holds cartridge 1 closed and allows cartridges 2 and 3 to open. Pump flow thus enters the annulus side of the cylinder via cartridge 3 and exhaust flow from the full bore passes back to tank through 2, causing the piston to retract.

Energising S2 allows cartridge 1 to open and connects pressure to the pilot ports of 2 and 3. Cartridge 2 will be held closed so pump flow passes freely across 1 into the full-bore side of the cylinder. Pressure applied to the full-bore side of the piston will create a pressure build-up in the annulus side and when this reaches system pressure plus spring pressure cartridge 3 will open, allowing the cylinder to extend in a regenerative manner.

A small amount of flow will pass from the annulus side back to tank via check valve A, but this will be limited by orifice B. When the piston reaches the end of its stroke, however, pressure in the annulus line will decay via orifice B, thus relieving the back pressure on the cylinder and giving full force from the piston.

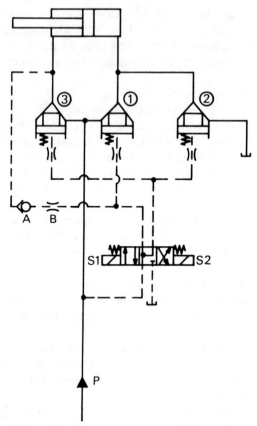

FIG. 45.   Regenerative circuit with full force at end of stroke.

## 5.3   Regenerative Circuit with Changeover to Conventional Circuit

With all solenoids de-energised (Fig. 46) all four cartridges are held closed, preventing cylinder movement.

Energising S1 allows cartridges 2 and 3 to open while 1 and 4 are held closed. Pump flow therefore passes across 3 to the annulus side and exhaust flow from the full bore is directed to tank via 2, causing the piston to retract.

Energising S2 allows 1 to open and holds 2 closed. Pressure will again build up in the annulus side which will open cartridge 3, giving a regenerative advance.

If solenoid S3 is now energised, this will allow cartridge 4 to open and allow the annulus flow to tank. Cartridge 3 will now close and the cylinder

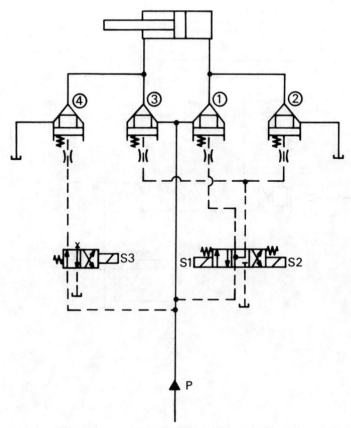

FIG. 46.  Regenerative circuit with changeover to conventional circuit.

will advance in the conventional manner, i.e. pump flow to the full bore and
annulus flow to tank.

### 5.4  Regenerative Circuit with Changeover by Pressure Sequence
This circuit (Fig. 47) operates in a similar way to that shown in Fig. 46
except that the changeover to a conventional circuit is achieved by means of
an unloading valve.

As the cylinder extends on light load, unloading valve A will remain
closed. The pressure in the annulus line will therefore hold cartridge 4
closed by means of the pilot through orifice B. Cartridge 3 will open and the
cylinder will extend regeneratively.

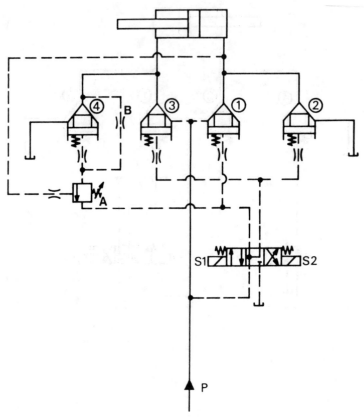

Fig. 47.    Regenerative circuit with changeover by pressure sequence.

When the load, and hence system pressure, have increased sufficiently, unloading valve A will open, thus opening the pilot connection of cartridge 4 to drain. This allows cartridge 4 to open and the cylinder to revert to a conventional advance.

### 5.5  Four-way Circuit with Anti-intensification Relief

When an anti-intensification or braking circuit is required, the standard 'four-way' circuit described earlier can be modified by changing one of the cartridges to a relief valve assembly (see Fig. 48).

To extend the cylinder solenoid S2 is energised, which allows valves 1 and 3 to open. If S2 is now de-energised while the cylinder is travelling at full speed, pressure will build up in the annulus side as the load decelerates. This

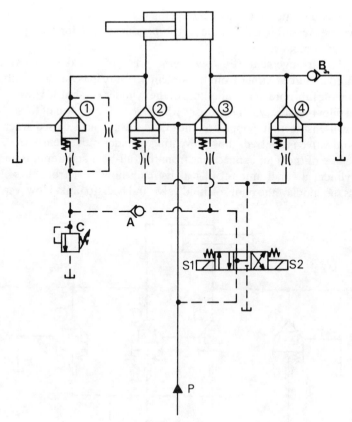

Fig. 48. Four-way circuit with anti-intensification relief.

pressure is now limited, however, by the setting of pilot relief valve C. Cartridge 1 thus applies braking pressure to the cylinder to bring it to a controlled stop, and check valve B allows make-up fluid to the full-bore side as the cylinder decelerates.

Check valve A prevents excessive flow to the pilot relief valve from the main pressure line.

Pressure in the annulus side of the cylinder is also limited by relief valve C during the retract stroke.

## 5.6 'Blocked Centre' Circuit

As mentioned previously, with a conventional four-way cartridge arrangement a reactive load on the cylinder when the solenoid valve is centered could cause it to move by forcing open poppets 2 or 3. Obviously this will be even more likely if the system pressure is lost, for example when pumps are unloaded.

In this arrangement (Fig. 49) pilot pressure is taken from three alternative sources: either from the system pressure line via check valve A, or from the full-bore or annulus lines of the cylinder via shuttle valve B. The actual pilot pressure will therefore be whichever is the highest of these three. This will ensure that poppets 2 and 3 are always held closed when the solenoid valve is centred, thus preventing cylinder movement.

Because of the pilot connections from the full-bore and annulus lines of the cylinder a small amount of leakage may now exist from these lines across the spool of the pilot solenoid valve and back to tank. However, this

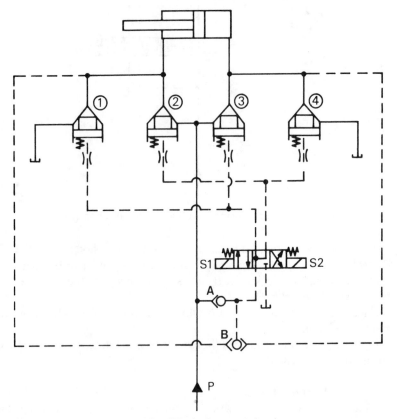

FIG. 49. 'Blocked centre' circuit.

leakage will normally be very small in relation to the main system flow rates and volumes.

### 5.7 Meter-in Circuit

By adding stroke adjusters to cartridges 2 and 3 of the standard four-way circuit (see Fig. 50) meter-in flow control of the cylinder can be achieved in both directions of movement.

The stroke adjusters and metering poppets will only provide non-compensated flow control, but the addition of a hydrostat will create pressure compensation if required.

### 5.8 Meter-out Circuit

As discussed earlier, if a restrictor poppet is used for meter-out flow control

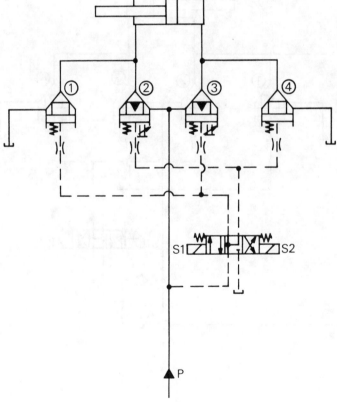

Fig. 50.   Meter-in circuit.

a problem may occur on the extension stroke of the cylinder. The intensified pressure created in the annulus side of the cylinder would tend to open cartridge 2 and create a regenerative circuit.

This can be overcome (see Fig. 51) by taking an alternative pilot supply from the annulus side through check valve A. The intensified pressure may then be used to hold cartridge 2 closed.

The system should be checked to ensure that the intensified pressure cannot reach dangerous levels when this arrangement is used.

### 5.9 Meter-out Circuit using 1·1:1 Poppet
In certain applications an alternative to Fig. 51 for meter-out applications would be to use a 1·1:1 poppet in position 2, as shown (see Fig. 52).

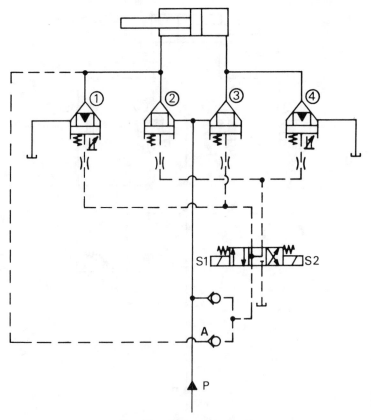

FIG. 51.   Meter-out circuit.

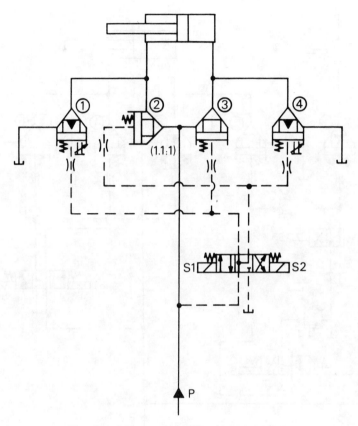

FIG. 52. Meter-out circuit using 1·1:1 poppet.

The intensified pressure now acts over only approximately 10% of the poppet area of cartridge 2. If a heavy spring is used in cartridge 2 (2·5 bar spring pressure), then the poppet will not open until the intensified pressure exceeds the system pressure by 25 bar.

This option may therefore be used in applications where the intensified pressure cannot exceed system pressure by more than approximately 25 bar, i.e. in cases where the rod area of the cylinder is small relative to the piston area.

## 5.10 Injection Moulding Machine Injection and Screw Control

Figure 53 illustrates a system applied to a plastic injection moulding machine. Following through a machine cycle, the sequence is as follows.

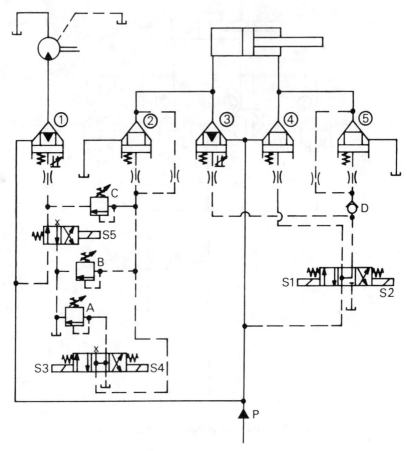

Fig. 53.  Injection moulding machine injection and screw control.

Solenoids S1 and S3 are energised initially. S1 allows cartridges 3 and 5 to open and the injection cylinder extends at a speed determined by the setting of restrictor poppet 3. The pilot connection of cartridge 2 is at this time connected to pilot relief valves B and C, and since the outlet of C is pressurised valve B determines the maximum injection pressure in the cylinder.

Solenoid S4 is then energised, now connecting the pilot port of cartridge 2 to relief valve A which therefore sets the injection hold pressure.

After a certain time solenoid S5 is energised, which allows cartridge 1 to open and the screw motor to rotate at a speed determined by the adjustment of poppet 2. This also opens up the outlet port of relief valve C which therefore sets cartridge 2 to the screwback pressure.

As the screw motor rotates the injection cylinder is mechanically pushed back (against the back pressure created by cartridge 2), with oil being drawn from tank over cartridge 5 into the annulus side of the cylinder (a slight positive pressure may be required in the tank line in order to overcome the spring in cartridge 5).

Solenoid S2 allows the injection cylinder to be hydraulically retracted if required.

## 6 INJECTION MOULDING MACHINE DESIGN

### 6.1 Basic Requirements
The basic requirement for all injection moulders is to increase productivity whilst maintaining the desired quality level of product. Increased productivity can be achieved by:

(1)  a shorter cycle time;
(2)  a shorter set-up time;
(3)  improving the quality of mouldings (i.e. a reduction in scrap levels);
(4)  optimising the energy consumption;
(5)  improving the environmental conditions (i.e. space, noise, etc.);
(6)  improving repeatability of parameters for the machine (i.e. flow, pressure, position, time, temperature) by processor control with suitable electro-hydraulic components;
(7)  adding peripheral units (robots, hoppers, dryers, etc.);
(8)  reducing storage costs as a result of smaller production runs;
(9)  reducing manpower with increased automation;
(10) improving moulds by better design.

### 6.2 Hydraulic Systems on Modern Injection Moulding Machines
The requirements of the oil supply systems on a modern injection moulding machine are as follows:

(1)  low noise;
(2)  optimised energy consumption—system efficiency;
(3)  high dynamic performance;
(4)  reliability.

Oil supply systems for injection moulding machines (IMM) for general production purposes are mostly a compromise of these requirements.

### 6.2.1 Small Machines

Small machines up to 120 tons clamping force are normally supplied by single pumps with proportional pressure and flow control (referred to as P/Q control) with fixed or variable displacement.

Figure 54 shows a variable-displacement piston pump with a load-sensing control and fitted manifold block with proportional pressure and flow control. The back pressure is controlled by the same proportional pressure pilot and will be selected by a solenoid valve. Due to the low dynamic characteristic of the pump a pilot-operated pressure relief valve has to be added for better decompression control and to avoid pressure peaks in the system.

The production of thin-wall mouldings and profiling of injection speed require a high dynamic performance of pressure and flow, therefore very

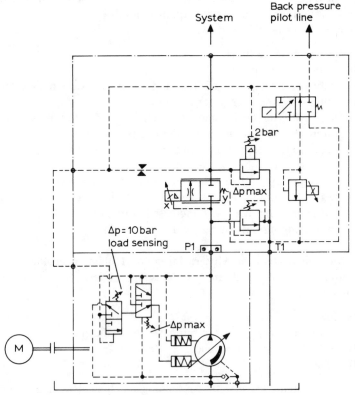

FIG. 54. P/Q control of variable pump with load sensing.

often the variable displacement feature has to be blocked and transformed into 'fixed displacement' during injection control.

Figure 55 shows a variable pump with a more effective control: the fitted manifold block with the proportional throttle and pressure valve has been replaced by a high dynamic P/Q spool valve. Pressure and flow are controlled by one valve only and the dynamic performance allows control in a closed loop.

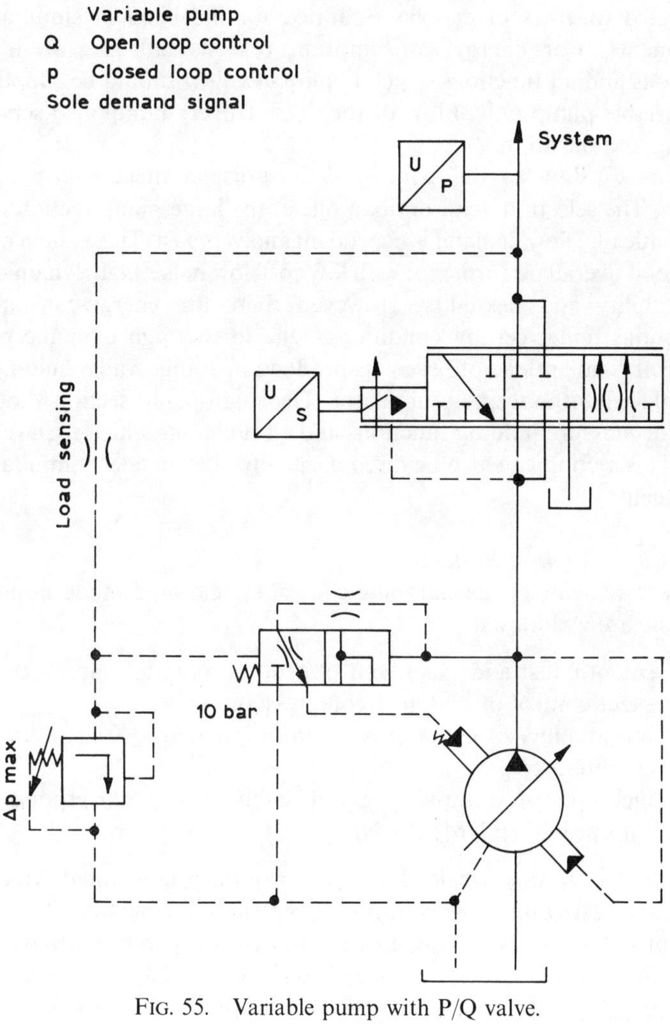

Variable pump
Q   Open loop control
p   Closed loop control
Sole demand signal

FIG. 55.   Variable pump with P/Q valve.

## 6.2.2 Medium-sized Machines

Medium-sized machines up to 400 tons clamping force are normally supplied by P/Q-controlled multiple pump systems with fixed and/or variable displacement. For the production of thin-wall mouldings very often an accumulator is added. Simultaneous movements are standard, therefore the selection of two independent circuits is required. Figure 56 shows a multiple pump system with variable plus fixed displacement. Each pump is P/Q-controllable. Both pumps can be connected together or can be separated for independent simultaneous movements. For energy consumption reasons, all pressure-holding functions and all functions which require low flow should be supplied by the variable pump only. Flow of the fixed delivery pump also serves for cooling and filtration.

Figure 57 shows a P/Q-controlled fixed-displacement multiple pump system. The selection 'load on/load off' of the large pump section occurs automatically; flow demand is dependent and very fast. This system offers a very good overall performance with low cost, low noise, high dynamic, high repeatability and flexibility. However, there are energy-consumption restrictions under certain conditions. Due to the high dynamic performance of 'load on/load off' even the production of thin-wall mouldings can be achieved without an accumulator. The small-pump section should be used for pressure-holding functions and all functions which require a low flow. This section can also be disconnected for independent simultaneous movements.

### 6.2.3 Clamp Control Systems

The requirements of the clamping control system on a modern injection machine are as follows:

(1) smooth, fast and exact control of force and motion;
(2) exact control of tie-bar decompression;
(3) repeatability of the stop position to less than $\pm 0.5$ mm under all conditions;
(4) inclusion of a moving guard safety device corresponding to European standards (CEN).

Figure 58 shows a toggle clamp control with proportional directional valve and selectable regenerative clamp-closing function. The safety element is a two-way cartridge valve with electrical position monitoring controlled by a NG6 roller-operated directional valve. The response time of the proportional directional valve should be within the range 30–45 ms to

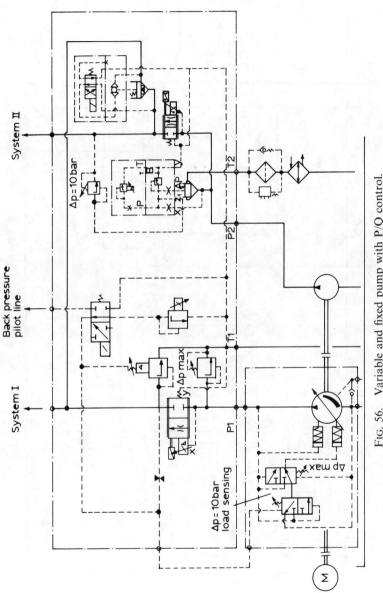

FIG. 56. Variable and fixed pump with P/Q control.

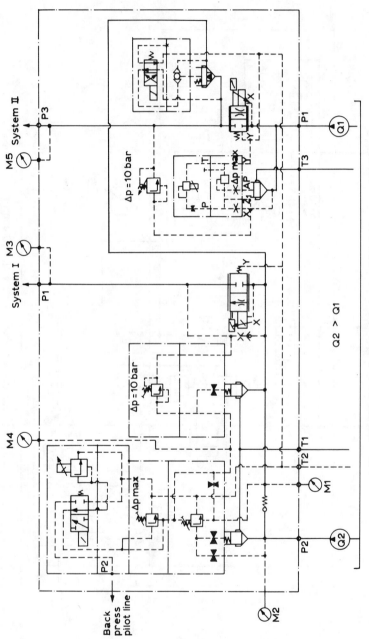

FIG. 57. Multiple fixed pump system with P/Q control.

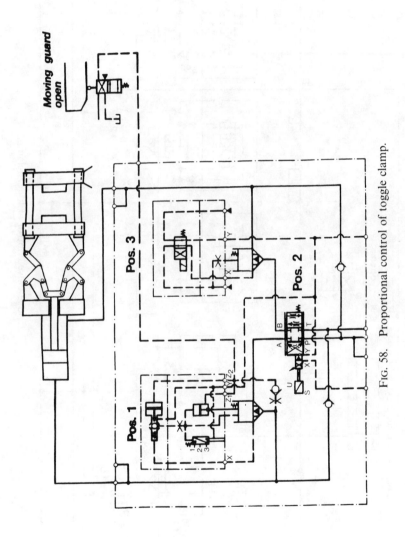

Fig. 58. Proportional control of toggle clamp.

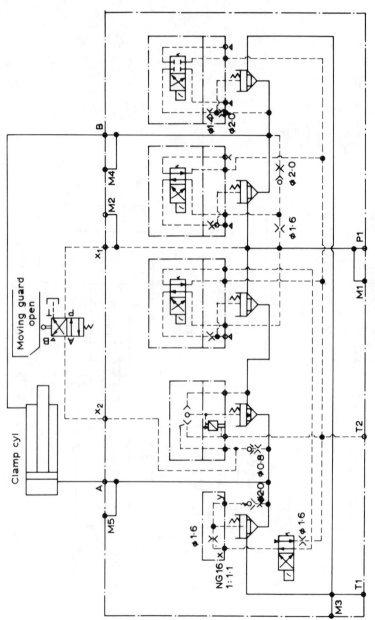

FIG. 59.    Cartridge valve control of toggle clamp.

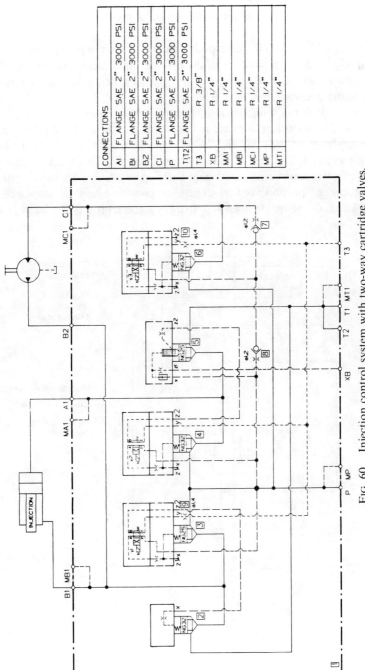

| CONNECTIONS | | |
|---|---|---|
| A1 | FLANGE SAE 2" | 3000 PSI |
| B1 | FLANGE SAE 2" | 3000 PSI |
| B2 | FLANGE SAE 2" | 3000 PSI |
| C1 | FLANGE SAE 2" | 3000 PSI |
| P | FLANGE SAE 2" | 3000 PSI |
| T1,T2 | FLANGE SAE 2" | 3000 PSI |
| T3 | R 3/8" | |
| XB | R 1/4" | |
| MA1 | R 1/4" | |
| MB1 | R 1/4" | |
| MC1 | R 1/4" | |
| MP | R 1/4" | |
| MT1 | R 1/4" | |

FIG. 60. Injection control system with two-way cartridge valves.

allow closed-loop deceleration-position control. Unsymmetrical spools are normally used corresponding to the cylinder area ratio. The regenerative function speeds up the clamp closing time and helps to achieve mould-protecting pressure (less than 5 bar).

Figure 59 shows a toggle clamp control with two-way cartridge valves. The speed is controlled by meter-in flow control of the P/Q unit of the pump system, therefore the mass and the decompression of the tie-bars at clamp opening have to be controlled by counter-pressure. This example shows a very effective and easily adjustable cartridge system with serial orifices, which cause a progressive flow-dependent counter-pressure characteristic. Oscillation of the spring mass system during acceleration–deceleration will

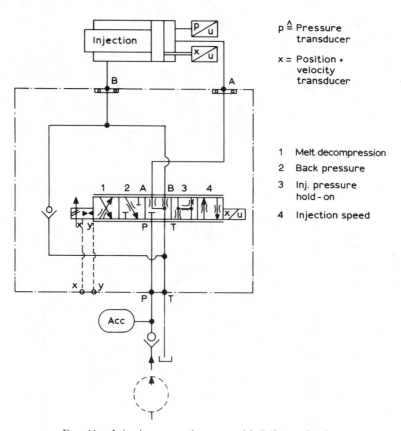

FIG. 61.   Injection control system with P/Q spool valve.

also be reduced by this counter-pressure characteristic and, as a result of this, a reduced clamp closing–opening time can be achieved.

### 6.2.4 Injection Control Systems

The requirements of injection control systems are:

(1)  exact and fast control of parameters with high repeatability, e.g. pressure should be controllable to better than $\pm 0.3$ bar, flow to better than $\pm 0.4\%$ and small signal resolution should be better than $0.1\%$;

(2)  high dynamic characteristic of controls and oil supply systems;

(3)  back pressure control down to a minimum of 1 bar;

(4)  profiling of injection speed and hold-on pressure.

Figure 60 shows an injection control system with two-way cartridge valves. Flow and pressure are controlled by the P/Q unit of the respective pump system. The special back-pressure cartridge consists of different pilot areas and will be piloted by the main proportional pressure valve of the P/Q unit. With a 1:4 area ratio, for example, back pressure can be controlled from 1 bar upwards and with the same signal resolution as the system pressure. The dynamic performance of the proportional flow and pressure valve should allow a closed-loop control.

Figure 61 shows injection control with a high dynamic P/Q-spool valve. Direction, speed and pressure can be controlled by one valve only. Speed and pressure are normally controlled in a closed loop. Because of the high pressure drop this valve is very suitable for all systems where oil is supplied by an accumulator.

# INDEX